Design of Fishways and Other Fish Facilities

Second Edition

Charles H. Clay, P. Eng.

CRC Press
Taylor & Francis Group
Boca Raton London New York

CRC Press is an imprint of the
Taylor & Francis Group, an **informa** business

First published 1995 by Lewis Publishers

Published 2019 by CRC Press
Taylor & Francis Group
6000 Broken Sound Parkway NW, Suite 300
Boca Raton, FL 33487-2742

©1995 by Taylor & Francis Group, LLC
CRC Press is an imprint of Taylor & Francis Group, an Informa business

First issued in paperback 2019

No claim to original U.S. Government works

ISBN-13: 978-0-367-44926-1 (pbk)
ISBN-13: 978-1-56670-111-2 (hbk)

Visit the Taylor & Francis Web site at
http://www.taylorandfrancis.com

and the CRC Press Web site at
http://www.crcpress.com

Published with the cooperation of the American Fisheries Society.

Library of Congress Card Number 94-17350

Library of Congress Cataloging-in-Publication Data

Clay, C. H. (Charles H.)
   Design of fishways and other fish facilities   /   by C.H. Clay. — 2nd ed.
        p.   cm.
   Includes bibliographical references and index.
   ISBN 1-56670-111-2
   1. Fishways—Design and construction.   2. Fish habitat improvement.
I. Title.
   SH157.85.F56C58   1995
   639.9′77—dc20   94-17350

AUTHOR

Since retirement from the Food & Agriculture Organization of the United Nations, Charles H. Clay has been a consultant to many organizations, companies, and governments throughout the world. He was born in New Westminster, British Columbia, and attended the University of British Columbia where he obtained a bachelors degree in 1944 in applied science in civil engineering. In 1962 to 1963, he completed the International Course in Hydraulic Engineering at Delft University in Holland, receiving a post-graduate diploma in hydraulic engineering. He has been a member of the Association of Professional Engineers of British Columbia since 1949 and is now a life member. His early work with the International Pacific Salmon Fisheries Commission saw him pioneering the work of salmon conservation in British Columbia and in 1950 this led to the position of Chief Engineer and Chief of Resource Development for the Federal Department of Fisheries in Vancouver, B.C. During this period much experience was gained in salmon conservation in British Columbia, the Yukon, and the western states of Washington, Oregon, and California. He also acted as adviser to the Canadian government on fish-migration problems on the east coast of Canada, Newfoundland, and the Great Lakes. With this accumulated experience, he wrote the first edition of *Design of Fishways and Other Fish Facilities* in 1961. Subsequently he continued to widen his experience of fish migration problems in other countries, including England, Scotland, Ireland, and The Netherlands, before joining the staff of the Food and Agriculture Organization (FAO) in Rome, Italy in 1965. With FAO, he served as Coordinator of Lake Projects, assessing and developing fisheries in the huge reservoirs formed by the new dams at Aswan in Egypt, Kainji in Nigeria, Volta in Ghana, and Kariba in Zambia, among others. He also has inspected projects and given advice on fisheries engineering problems in India, Iran, Poland, Greece, Iceland, and many other countries. He has attended conferences and presented papers to the International Union for the Conservation of Nature in Athens, Greece; the American Geophysical Union in Knoxville, Tennessee; and the International Symposium on Fishways at Gifu, Japan. Now he has collected all this knowledge, and with the help of workers in fisheries throughout the world, has compiled a second edition of *Design of Fishways and Other Fish Facilities* bringing the subject up-to-date from a world perspective. In 1993 he received an Award of Excellence from the Bioengineering Section of the American Fisheries Society.

PREFACE AND ACKNOWLEDGMENTS

This new edition was inspired by my ex-colleagues in the Fisheries Department of the United Nations Food and Agriculture Organization (FAO) in Rome. They were embarrassed, when receiving enquiries concerning fishways, by having to tell their correspondents that the original volume, the only one of its kind, was over 20 years old, and more recently, that it was out of print. I then entered into an agreement with Dr. Fran Henderson, Director of the Resources and Environment Division of FAO, to begin work on the new edition, provided it would be expanded as much as possible to include facilities for fish of interest to the developing countries.

During this initial phase, it soon became evident that there had been very little progress in defining the problem for the types of fish present in the developing countries, particularly those in the southern hemisphere. Nevertheless, I gathered what information I could on experience in these countries, in addition to updating all material on the developed countries of the northern hemisphere.

I am deeply indebted to all my correspondents for their assistance. In the United States, Milo Bell, George Eicher, and Ted Vande Sande were very helpful. In Canada, Chris Katopodis, Ray Biette, Vern Conrad, H. Jansen, and Paul Ruggles supplied valuable information. M. Larinier in France, M. Beach in England, C. McGrath in Ireland, T. Gudjonsson in Iceland, G. Jens in Germany, and L. Van Haasteren in The Netherlands were most helpful. Similarly, A. H. Bok in South Africa, J. H. Harris and M. Mallen-Cooper in Australia, C. Mitchell in New Zealand, V. Pantulu and H. R. Rabanal in Southeast Asia, and Xiangke Lu in China provided help by correspondence. The publications by D. S. Pavlov of Russia and R. Quiros of Latin America, sponsored by FAO, were most timely. This help from a large number of sources enabled me to give an up-to-date accounting of fishway problems and practices all over the world.

Funding arranged by the American Fisheries Society enabled me to complete the manuscript and the new illustrations; the late Carl Sullivan was instrumental in this regard. The following organizations generously provided financial support:

· Food and Agriculture Organization of the United Nations
· Canada Department of Fisheries and Oceans
· U.S. Department of the Interior, Bureau of Reclamation

Permission has been granted for the use of the figures from the original publication by the Canada Department of Fisheries and Oceans, Ottawa. J. Clay made the new diagrams with a computer, and J. Johnson typed the manuscript. To all of these people and organizations, a sincere thank you is offered for helping in this second edition.

Charles H. Clay

CONTENTS

Preface and Acknowledgments

1   FISHWAYS — GENERAL

1.1 INTRODUCTION

Many different types of devices have been used to facilitate migration of fish past dams, waterfalls, and rapids. The type of fish facility that will be described in this chapter is known in general as a *fishway, fish ladder,* or *fish pass*. The terms *fishway* and *fish ladder* are used in North America, and *fish pass* is used in Europe. For the sake of uniformity, *fishway* will be used throughout this text. Chapters 2 and 3 will deal with the types of fishway that enable the fish to swim upstream under their own effort. *Fish locks* and *fish elevators,* which lift the fish over an obstruction, will be dealt with in Chapter 4.

Regardless of what they may be called in different areas of the world, and the species of fish they are designed to accommodate, fishways all have basically the same definition. They are essentially a water passage around or through an obstruction, designed to dissipate the energy in the water in such a manner as to enable fish to ascend without undue stress.

Fishways have a long history, with the earliest ones recorded almost 300 years ago in Europe. Undoubtedly, there was a realization of the need for fishways even before this, but in those earlier times, as one would expect, the problems involved in meeting this need were very poorly understood. Unfortunately, almost the same lack of understanding has persisted down to modern times, and we are now in the position of having to overtake, in a matter of a few years, the fish-passage problems created by a hundred years of industrial development. In many areas it is too late to apply our knowledge, because the populations of migratory fish have been completely destroyed, and the problems of restoring them in such cases are often insurmountable. There are, however, many migratory species of fish left in the rivers of the world, and our growing confidence in scientific research and management, coupled with the recognition of the need for preservation, cannot help but enable us to achieve a large measure of success in preserving and even increasing their numbers. The construction of adequate fishways is one of the means needed to achieve this end.

1.2 SOME LEGAL CONSIDERATIONS

The awareness of the need for providing facilities for fish passage at obstacles in rivers appears to have grown simultaneously in most northern countries where salmon runs have occurred. Undoubtedly this awareness was a direct result of the obvious distress displayed by jumping fish at newly erected weirs. As the biological sciences progressed, the need was given public recognition by provisions for it incorporated in our modern laws.

Many countries and states with important fisheries dependent on populations of migratory fish now have specific laws and regulations protecting fish that are affected by dam construction and other water-use projects. In North America the laws are comparatively uniform in concept and administration. They provide that the owner of any dam or other obstruction to migration be responsible for providing facilities for fish passage. While the law places the onus for construction of a suitable fishway on the *owner* of the dam, the owner is not expected to be an expert in fishway design, so either the government or a consultant working closely with the government is obliged to give advice to the owner concerning the design that will meet with approval. This policy has led to the practice of providing the owner of the dam with what have been described as *functional* plans and specifications. These commonly include the general layout, interior dimensions, and specifications covering the hydraulic features of the fishway. The structural design of the fishway is done by the owner along with design of the dam, and the final plans are then approved by the responsible agency of the government.

In the text that follows, fishways will be discussed in a manner that is in keeping with this procedure. For fishways at dams, very few structural details will be described, because such information can be readily obtained from standard references dealing with the design of dams and other hydraulic structures. However, the functional and hydraulic requirements will be discussed in detail, since they are a responsibility of the fisheries scientist. In the discussion of fishways at natural obstructions, on the other hand, the whole problem will be considered as a unit, including the conception, survey, hydraulic and structural design, construction, and operation of the fishway, because in this case fisheries engineers and biologists are normally responsible for the entire project

In outlining the approach to the design of fishways at dams, it has been taken for granted that the cases under consideration are those where the public interest is best served by permitting construction of the dam. This may not always be the case. In some instances it may be apparent that the best facilities for fish passage cannot assure perpetuation of the fishery. If the fishery is very valuable, the public interest might be better served by not building the dam. In this text it is assumed in dealing with fish facilities at dams that it has been judged in the public interest to permit construction of the dam, on the condition that the best fish facilities that can be economically provided will be required.

1.3 DEFINITIONS

In order to define some of the terms used in describing various parts of fishways, it is necessary to know the basic types of fishway. The fishway that has probably been most commonly used throughout the world consists of a series of pools in steps leading from the river above a dam to the river below, with water flowing from pool to pool. The fish ascend by jumping or swimming upstream from pool to pool. The pools are separated by *weirs,* which are simply barriers or dams that control the water level in each pool. These are known as *pool* and *weir type* fishways or more simply as *weir type,* which is the designation that will be used in this text. While this type of fishway probably originated with a series of ponds or pools excavated in steps around a dam, nowadays it is usually constructed by dividing a flume of rectangular cross section into a series of compartments or pools by cross walls, which act as weirs. These cross walls are known as *baffles.*

While weirs have been frequently used to divide the fishway into pools, a cross wall with no weir flow over it has also been used. The entire flow between pools is passed through a submerged orifice, and the fish ascend from pool to pool through the orifices. These fishways are known as *pool and orifice* fishways, or simply *orifice* fishways, as they will be referred to in this text.

Most modern weir type fishways also have orifices in the baffle walls, but they are still referred to as weir type, provided that there is sufficient spill over the baffles for fish to swim over them.

With both the weir and orifice type of fishway, the energy given up by the water in passing from pool to pool is dissipated in turbulence in the pool itself. Ideally the flow is not large compared to the pool volume so that it is possible for the pool to absorb this turbulence and dissipate its energy before the water flows over or through the baffle to the next pool. In this way the energy is dissipated uniformly down the fishway, in a manner that permits the fish to ascend, ideally without undue stress.

In Belgium more than 80 years ago, a fisheries scientist named Denil conceived the idea of dissipating the energy of the water flowing down a steep flume or trough by installing vanes on the sides and bottom, thereby causing part of the flow to turn back on itself. The velocity of the water turned by the vanes met the velocity of the main stream in the central part of the fishway, and the result was a reduced velocity against which the fish were able to ascend. This type of fishway is known as the *Denil type,* and many variations of it have been devised, all of which will be classified as the Denil in this text.

A fourth type of fishway is known as the *pool and jet type.* This is similar to the pool and orifice type except that the orifice extends over the full height of the baffle, so that it is in the form of a vertical slot rather than a submerged orifice. The flow characteristics and hydraulic properties are therefore different from the orifice type. A particular version of the pool and jet type of fishway was developed about 1943, and is known as the *vertical slot* or *Hell's Gate type.* It incorporates to a certain extent the principle Denil previously exploited, and for that reason it is really a combination of types. It has been widely and successfully used for the passage of salmon and other species in North America and Europe. This fishway is constructed by installing a series of baffles at regular intervals between the walls of a flume. The baffles are so shaped as to partially turn the flow from the slots back upstream, with the result that if the slots are properly shaped and dimensioned, the energy dissipation is excellent over a wide range of levels and discharges. It has the added advantage of permitting the fish to swim through the slots from one pool to the next at any desired depth, since the slot extends from the top to the bottom of the flume. This design was developed under the supervision of Milo C. Bell specifically for use at Hell's Gate on the Fraser River in Canada, and for that reason it is often called the Hell's Gate type. However, because of its wide use at other locations, it will be referred to as the vertical slot fishway in this text.

Other terminology associated with fishways is usually derived from the concept that the fishway is a passage upstream for fish rather than a passage downstream for water. For example the downstream end is the *fish entrance* or more simply the *entrance*, while the upstream end is the *fish exit*, or simply the *exit*. This is often confusing to hydraulic engineers, who are more used to thinking in terms of the water entrance and water exit. However, the principle implicit in the use of the terms *fish entrance* and *fish exit* is significant, since it is necessary to regard all fishway problems from the point of view of fish passage.

Fish entrances are perhaps the most important parts of fishways, particularly at dams, and a special section later in this text is devoted to the many variations and requirements for them. The *powerhouse collection system*, described later, is merely an arrangement of fish entrances above the powerhouse draft tubes, so placed as to enable fish attracted to the outflow from the draft tubes to find an entrance as readily as possible.

In locations where the flow from a fishway is small compared to flow in the river, an *auxiliary water supply* may be required to increase the velocity of the flow out of the fish entrance and as a result attract fish more readily. This is often known as *attraction water*, and some of the presently used specifications for quantities and velocities are described in later sections.

Fishways are frequently located on or near river banks, where they are accessible to fish migrating up the margins of the river. Often there is a fishway on each bank at natural obstructions and at dams. To differentiate between fishways in such cases, it may be necessary to designate them as pertaining to one bank or another. When necessary, the standard method will be used in this text, and the fishways will be designated as *left bank* or *right bank* as they would appear to a person facing downstream.

Fish locks and *fish elevators* are dealt with in Chapter 4 of this text although in reality they are fishways of a different type. A *fish lock* is defined as a device that raises fish over dams by filling with water a chamber the fish have entered at tailwater level or from a short conventional fishway until the water surface in the lock reaches or comes sufficiently close to reservoir level to permit the fish to swim into the forebay or reservoir above the dam. It is similar to a navigation lock, and indeed, fish have been known to pass upstream through navigation locks on many occasions. A *fish elevator* is defined as any mechanical means of transporting fish upstream over a dam, such as a tank on rails, a tank truck, a bucket hung on a cable, etc., and will in this text include the means of collecting and loading the fish into the conveyance.

Figure 1.1 illustrates in simplified form the main types of fishways, fish locks, and elevators as defined here.

1.4 HISTORY

As already noted, the earliest recorded attempts to construct fishways were made in the 17th century, and undoubtedly there were earlier attempts of an even more

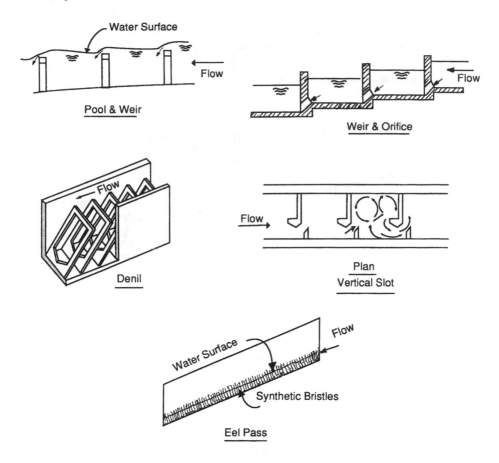

**Figure 1.1   Cross sections of some of the fishways and other methods of facili-
tating fish passage over dams. The vertical slot fishway is shown in
plan for clarity.**

primitive nature. No attempt appears to have been made to approach the problem on
a scientific basis until the 20th century, however, even though there had been
tremendous advances in the science of hydraulics and related fields in the meantime.

In his annotated bibliography of fishways, published in 1941, Nemenyi lists
many publications concerning fishways which appeared during the latter part of the
19th century and early part of the 20th century. None of these publications appeared
to have been based on a rational scientific approach, however, although some
included detailed plans of fishways that were fairly elaborate for the time.

In 1909 Denil published a paper that described the new type of fishway he had
developed, which was based on more scientific principles, and from that time to the
present there has been a distinctly noticeable change in the approach to fishway
design. First there appears to have developed a phase where the hydraulics of
fishways as energy dissipators were studied and became more clearly understood. In
addition there were attempts to understand the mechanics of swimming fish. These
latter studies appear to have been an offshoot of the engineering studies, and consequently they tended to regard the fish more as inanimate objects, subject to various

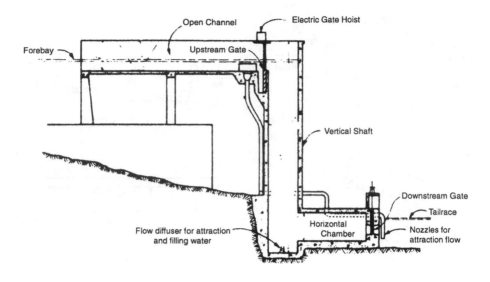

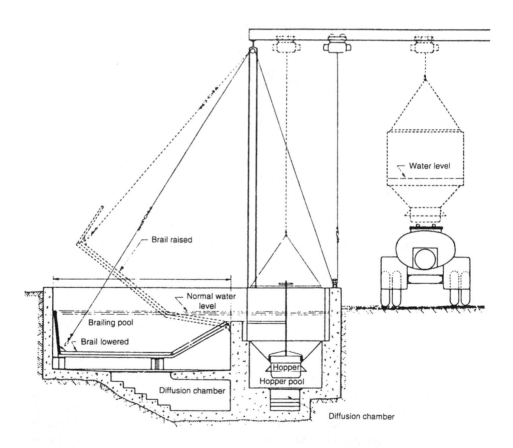

Figure 1.1 (continued). Fish lock (top) and fish elevator (bottom).

calculable or measurable external forces, than as living animals possessed with definite physiological limitations and behavior patterns.

Nevertheless, these studies by Denil and others did advance the science greatly, and the Denil fishway itself, as well as the principle it incorporates, have been widely used up to the present time. While Denil and other investigators recorded certain observations on fish behavior incidental to their work, it remained for McLeod and Nemenyi to undertake some research in 1939 and 1940 involving the actual performance of fish in relation to a number of different types of fishways. This was a decidedly empirical approach, but was still of considerable value even though it was limited to fish that were native to the streams of Iowa.

Meanwhile, of course, dams were being built with fishways included, and fishways were also being designed to overcome natural obstructions in accordance with the best knowledge available. From these installations considerable experience was being gained relating to the behavior and abilities of the fish involved, but because of the haphazard approach, progress was slow. However a giant step forward, particularly as it applied to salmonids, was realized with the design and construction of the fish facilities at Bonneville Dam on the Columbia River in 1937 to 1938, shown in part in Figure 1.2. Here, for perhaps the first time, sufficiently large numbers of salmon were involved to warrant large expenditures on designs incorporating the latest ideas of experienced engineers and fishery biologists. As a result, the fish facilities for passing adult salmon upstream were larger, more complex, and probably more efficient than any previously constructed. The design incorporated new principles, which included the use of large quantities of attraction water and a multiple-entrance powerhouse collection system. Fish locks were also included along with standard weir type fishways, and provision for downstream migrant juvenile fish was attempted. While the different components met with varying success, the adult facilities in particular were so advanced that much of their general design is still accepted as a standard on many installations in a wide range of locations.

The U.S. Fish and Wildlife Service and the Departments of Fisheries of the States of Washington and Oregon were the agencies responsible for the hydraulic and biological details of these fishways, and to their biologists and engineers must go the credit for this tremendous advance.

Shortly after this, the International Pacific Salmon Fisheries Commission, which had been making a thorough biological study of the Fraser River sockeye salmon, recommended relieving the obstruction at Hell's Gate Canyon on the Fraser River in Canada by the construction of fishways.

The vertical slot baffle was evolved to meet the particular conditions at Hell's Gate (Figure 1.3), and this development represented another great advance in fishway design. The principle utilized, as pointed out previously, consists of turning the flow from a jet back on itself, thus increasing the efficiency of energy dissipation in the pools. In addition, the geometric pattern of eddies that were formed, which was reasonably stable at all depths, provided better resting areas. Because of its ability to accommodate itself to wide fluctuations of headwater and tailwater elevations, while still maintaining simplicity of design at a reasonable cost, this type of fishway has proved to be ideal for installation at natural obstructions.

Over the half century since construction of the Bonneville facilities, there has been an increasing awareness of the complex nature of fish-passage problems, and

Figure 1.2 Bonneville Dam with Bradford Island Fishway in the foreground.

Figure 1.3 The Hell's Gate Fishways, looking upstream.

this has led to a great deal of basic research on the problem of downstream passage over dams as well as a questioning of the efficiency of upstream facilities.

Proposals to build many more dams on the Columbia River were the main stimulant to an assessment of existing facilities and increased research aimed at improving new facilities. Two separate investigations followed.

The first, an investigation of losses to downstream migrants over Bonneville Dam as reported by the U.S. Fish and Wildlife Service (1948), revealed that these losses were 15% overall. The second, an investigation by the Oregon Fish Commission reported by Schoning and Johnson (1956), revealed that a delay averaging several days occurred in upstream migration through the fish facilities at Bonneville Dam.

These investigations emphasized the need for further research to improve upstream fish facilities and to eliminate downstream mortality as much as possible.

Long-term programs of basic biological research on the physiology and behavior of migrating salmonids were undertaken with the hope that a complete understanding of the biological requirements would eventually enable engineers and biologists to design fish facilities at dams that would fully protect the valuable salmon runs. The need for basic biological knowledge of fish in the fields of physiology and behavior has been recognized by other countries, and several of them have their own programs with similar objectives. This is not considered to be a duplication of effort, because it is obvious that the requirements are variable. The more effort expended on different species in a number of different locations, the better. Not only will the understanding of the particular locations be improved, but it will also be possible to generalize with more confidence over wider areas.

Both fish locks and fish elevators have a shorter history than conventional fishways. A Mr. Malloch, of Perth, Scotland, has been credited with evolving a scheme similar to modern fish locks around 1900, but evidently this was too early for the idea to be accepted, as no immediate applications followed. The period of the initial application of fish locks and elevators on a practical scale corresponds to a period when dams started to be designed that were much higher than any previously conceived. This was in the mid-1920s. As long as dams were less than 50 ft (16 m) in height, conventional fishways were not considered to be unduly expensive. As dams higher than this became more common, and some were envisaged on salmon streams to heights of 300 ft (100 m), there was more incentive to find alternative methods for providing adult fish passage. Another factor besides economy was instrumental in the development of these devices. This was the fear that fish would not be physically capable of ascending fishways over such dams. More is being learned each year, however, and fishways have been constructed at dams of increasingly greater heights so that the earlier fears have been reduced to some extent.

The lack of data on the stresses imposed on the fish by their ascent through fishways at dams of varying heights has undoubtedly been instrumental in the development of both fish locks and fish elevators, because both of these devices require less effort on the part of the fish than conventional fishways. This does not say that they are recommended, however, because there is also a lack of data on the stresses imposed on the fish by passage through these devices, so that the advantage of less effort is more apparent than real.

In his bibliography, Nemenyi (1941) refers to articles describing an experimental fish elevator tested on the White Salmon River in the western U.S. about 1924, and to a patented fish lift or lock tried on the Umpgua River in Oregon in 1926. The earliest similar installation in Europe is noted in the same bibliography as an elevator at Aborrfors, Finland, first described in 1933. This was followed, apparently within a few years, by several other installations in Finland, and by an elevator on the Rhine River at Kembs in Germany.

About this time, interest in North America turned to an installation at Baker Dam in the State of Washington, which utilized a cable-and-bucket system to pass fish over a total height of almost 300 ft. This contrivance was hailed at the time as the answer to the problem of passing fish over all high dams, but it is interesting to note that it was replaced by a new trapping and trucking operation. The decline of the salmon runs to the Baker River in the intervening period of about 30 years was not attributed wholly to the effects of the original hoist, because of the high mortality to downstream migrant juvenile fish at the dam, which was measured and recorded by Hamilton and Andrew (1954). However, the fact that the system for adult fish passage was replaced at considerable cost is assumed to be evidence of dissatisfaction with the original method. Bonneville Dam, which was also constructed in the 1930s, was furnished with large fish locks in addition to the elaborate conventional fishways described previously. These fish locks were mainly experimental, but have served a useful purpose as well, as will be described later.

There are no publications to indicate that there was further construction of fish locks or elevators on any extensive scale until the period following the end of World War II, 15 to 20 years later. This does not mean that there was a complete cessation of interest in these devices in the interim, however. In the period 1939 to 1943, following construction of Grand Coulee Dam, a temporary trapping and trucking operation at Rock Island Dam on the Columbia River some distance downstream, was successfully used to transport several thousand adult salmon from the latter dam to new spawning grounds. Fish and Hanavan (1948) report the details of this operation, which was necessitated by the loss of the original spawning grounds above Grand Coulee.

After World War II, development of fish locks as practical fish-passage facilities accelerated in Europe, starting with the Leixlip development on the River Liffey near Dublin, Ireland, about 1950, and continuing in Scotland and Ireland up to the present time. Also during the following years a trend to this type of installation in Holland was recorded by Deelder (1958).

In the same period fish elevators of the trapping and trucking variety were developed as practical facilities for fish passage at high dams in the U.S. and Canada. The White River trucking operation, to provide fish passage over the Mud Mountain Dam in Washington State, was one of these.

In addition, trapping and trucking operations were used on the Sacramento River in California in the years following 1943. Moffett (1949) describes the results of these operations at Keswick and Balls Ferry below Shasta Dam. The new Columbia River dams at McNary and The Dalles were furnished with fish locks in addition to conventional fishways. These locks were partly for practical and partly for experimental purposes and cannot be placed in the same category as the European fish locks, because they do not furnish the main or only means of ascent for fish at either of the dams.

There have only been sporadic installations of fish locks in North America in the last 30 years. First there was one on the Connecticut River near Holyoke, MA. This was intended to pass American shad upstream, but was found to be unsatisfactory, and has since been replaced by an elevator consisting of a bucket or hopper, which is used to transport the fish over the dam. This has proved to be highly successful and has served as a model for construction of a hoist facility in 1980 at Essex Dam on the Merrimack River in Massachussetts. A similar hoist facility was also constructed at the Mactaquac Dam in New Brunswick on the St. John River in 1967.

A fish lock installed at Thornberry on the Haines River in Ontario is reported to pass large numbers of rainbow trout and chinook salmon from the Great Lakes over a dam 7.3 m (23 ft) high. A lock was installed at a small dam on Brunette Creek in Western Canada but has not operated very satisfactorily.

Meanwhile many Russian installations were constructed after World War II. S. M. Kipper (1959) describes a lock that has a main shaft 8.5 by 8.5 m, and two approach channels. He also notes a fish elevator installed on the Don River at Tsimliansk. Klykov (1958), Z. M. Kipper and Mileiko (1967), and Pavlov (1989) describe many of these installations to transport acipenserid and cyprinid fishes over dams.

To summarize, modern fishway, fish lock, and fish elevator design has evolved over a period of several hundred years, with the most rapid advances occurring in the last 50 years. Even during this latter period, however, improvements have been slow because of a lack of data on the biological requirements of the fish. It is believed that this weakness will be remedied in coming years and that fish facilities will be greatly improved as a result.

1.5 CURRENT USE OF FISHWAYS, LOCKS, AND ELEVATORS AROUND THE WORLD

It is appropriate to review now the current status of the use of fishways, fish locks, and fish elevators around the world. This review is useful for several reasons. It will show the wide experience of differing types of facilities in different countries. It will show how many different species are being dealt with and will enable the reader to get a preliminary idea of the type to consider for his or her own particular problem. The record of experience with the species of fish the facility must be designed to pass will in many cases be an important factor in the design of new facilities. If there is a long record of fishways of one of the types noted having successfully passed runs of the species under consideration, one would be reluctant to risk experimenting with a different type, particularly if the fishery that will be affected is of some value. The general picture of experience with fish facilities throughout the world, as far as the writer is aware at present, is as presented below.

1.5.1 North America

1.5.1.1 West Coast

On the Pacific Coast of North America, weir-type fishways with orifices through the baffles have a long history of use at dams, and much first-hand information is available from the experience gained at the many large dams on the Columbia River. Most of these fishways are operated with 1 ft (30 cm) of head per baffle. These have gradually been refined over the years. One of the more recent installations, at Ice Harbor on the Snake River near its confluence with the Columbia, includes a number of orifice baffles at the upstream end of one fishway, with the balance of the baffles being of the weir type. The weir type with orifices in the baffles has been utilized by all five species of Pacific salmon, by steelhead, and by American shad, common carp, northern squawfish, sturgeon, suckers, lamprey, shiner, whitefish, chub, dace, bass, crappie, catfish, rainbow smelt, and smaller trout. While this would indicate that this type of baffle might be considered for use with any of these fish, it must be remembered that these fishways were designed primarily to pass Pacific salmon and steelhead trout and not specifically to pass any of the other species mentioned. However, because of the existence of orifices in the baffles (which eliminates the need for the fish to jump or swim over the weir), there is every reason to believe that this type of baffle could be used by almost any migratory fish, provided the head between pools was adjusted so that water velocities were within the swimming capabilities of the fish and no excessive turbulence occurred in the pools.

The vertical slot baffle has been used to overcome natural obstructions in streams along the Pacific Coast. Because of its advantage in requiring no manual regulation of flow under the condition where variations of head- and tailwater are reasonably comparable, it was bound to be used at dams where this condition is fulfilled. This type of fishway was installed at a concrete hydroelectric diversion dam on Seton Creek and at a concrete storage dam on Great Central Lake, both in British Columbia. It has also been used for temporary fishways around the cofferdams during construction of some of the Columbia River dams. This experience with vertical slot fishways at both dams and natural obstructions provides considerable confidence in their application at any installation where hydraulic conditions are favorable. The existing installations of this type have successfully passed all five species of Pacific salmon, steelhead, and small cutthroat and rainbow trout. Large numbers of lampreys have also been observed migrating through them. Because this baffle permits fish to ascend at any depth they choose, it is also considered to be suitable for any migratory species of fish, provided the head conditions between pools are adjusted so that velocity and turbulence are within the swimming ability of the fish.

A weir type fishway was installed in a dam on the Yukon River near Whitehorse and came into operation in 1959. The most northerly fishway installation of its kind in Canada, it was reported to have successfully passed Pacific salmon (chinook), Arctic grayling, lake trout, least cisco, whitefish, and suckers over the 50-ft (16-m) height of the dam. Later studies, done by Cleugh and Russell in 1980, found the salmon were being delayed in entry to the fishway although there was apparently no indication of delay in ascending over the weirs once they had entered.

The Denil fishway has been tested and used on the Pacific Coast of North America to pass sockeye, chinook, and coho salmon. An experimental installation was made at Dryden Dam on the Wenatchee River in the state of Washington, and was thoroughly tested as reported by Fulton et al. (1953). These tests indicated that two species of Pacific salmon (chinook and sockeye) would readily ascend a well-designed Denil fishway, as would other species such as rainbow trout, Dolly Varden, suckers, and squawfish. This Denil fishway was based on a design recommended by the Committee on Fish Passes, British Institution of Civil Engineers (1942). The design was modified in line with changes described by Furoskog (1945) for a fishway at Hurting Power Dam in Sweden which increased the linear dimensions of the fishway by about 42%.

The Denil fishway is not widely used, however, except in Alaska, where its light weight and mobility when constructed of aluminum have proved invaluable for placement at natural obstructions that are inaccessible except by helicopter. It is described by Zeimer (1962).

Experimentation has been done recently with a new type of baffle for a pool and weir fishway for Pacific salmon, which requires the fish to jump over the baffle, thus reducing the number of baffles in the fishway by increasing the height of each baffle. Orsborn (1985) has reported on this work. The work started as a detailed analysis of what determines fish behavior when it reaches an obstruction during its migration, and then attempts to answer the question proposing a baffle design that causes the fish to exert itself to near its limit of effort. While the report proposes as a result a baffle with a 2.5-ft drop in head, and claims that in field tests all coho and chum salmon surmounted this baffle, it remains the subject of a great deal of skepticism among fishery biologists and engineers. Until the advantages of this baffle have been thoroughly evaluated in the field, the outcome of the work remains in doubt.

Fish locks have been tested on the Pacific Coast of North America but have not been found to be practical for conditions there. Fish elevators have been tested and used extensively at many of the higher dams. They have been found to be satisfactory for all species occurring in the area, providing good design principles have been followed.

1.5.1.2 Central

There are a number of fishway installations in Central Canada and the U.S., mainly at water-control dams. These are mostly pool and weir type, showing varying degrees of success, particularly with respect to passage of northern pike and walleye. Several different types of fishways were experimented with, but many failed to perform satisfactorily, chiefly due to entrance conditions, which were adversely affected by large fluctuations in flow.

In Canada, 35 fishways are reported to exist in the provinces of Alberta, Saskatchewan, and Manitoba, almost all of the pool and weir type (Washburn & Gillis Assoc., Ltd., 1985). Fish species passed are northern pike, walleye, cisco, brown trout, Arctic grayling, and mountain whitefish. Research at the University of Alberta has demonstrated an awakening of interest in the Denil fishway, as reported by Katopodis and Rajaratnam (1983). The hope is that the Denil can answer some of the entrance problems.

In the central states, five Denil fish passes were reported at dams in the Des Moines River (Katopodis and Rajaratnam, 1983). These and other fishways in Iowa were reported to pass all species over 15.2 cm in length, the majority of which were catostomids, cyprinids, and ictalurids. Also included were large northern pike and walleye.

1.5.1.3 Great Lakes Area

This area includes Ontario in Canada and the U.S. states that contain tributaries of the Great Lakes, such as Wisconsin, Illinois, and Indiana, to name a few. A program of planting Pacific salmon in the Great Lakes has been so successful that it has emphasized and in some cases created a need for fishways in many of the rivers tributary to the lakes. The principal species present, in addition to chinook and coho salmon, is the rainbow trout, which migrate upstream in the spring. They grow to about 7 lb and migrate in large numbers. Walleye are also present as are lampreys, and this unique mixture of species requires special treatment at dams.

The lowermost dam in a given river system is usually designed as a lamprey barrier, in an attempt to eliminate this species. This has been a continuing effort for at least 30 years, since the lamprey first gained access to the lakes through the shipping canals and locks and started to prey heavily on the indigenous trout population. The barrier consists of an overhanging lip on the first weir, extending for about 8 in. (20 cm). On most fishways (mainly weir type) this is effective in excluding the lamprey while allowing the trout and salmon to jump over the baffle.

In Ontario a lock of the Borland type (see Chapter 4, Figure 4.1) has been constructed and operated successfully, passing both rainbow trout and chinook salmon.

1.5.1.4 *East Coast*

On the Atlantic Coast of the U.S. and Canada, fishways of all types have been used to pass Atlantic salmon, shad, American eels, alewives, trout, and striped bass.

In the maritime provinces of Canada, fishways of the pool and weir type predominate, although in recent years there is a commendable trend to use the fishway best suited to the conditions encountered, whether it be weir, vertical slot, or Denil type. Fish lifts have also been used at relatively high dams on the St. John River and tributaries. Denil fishways are less widely used than they are in New England and Europe.

Conrad and Jansen (1987) report that in this area there are 200 fishways of the pool and weir type, 25 vertical slot, and 15 Denils. For salmon only, they use drops between the pools of up to 2 ft. Alewives and possibly shad must have a lower head per pool (9 in., or 23 cm) and a special baffle, which will be described later in this text. They do not pass readily through a small vertical slot fishway.

In recent years the Denil fishway has been used widely in Maine. A design using a single plane baffle, which was based on the Committee on Fish Passes (1942), is favored, with a slope of 1:5 (vertical to horizontal). One Denil fishway in Maine is 227 m long (including resting pools) and surmounts a vertical height of 15.2 m (Decker, 1967). Katopodis and Rajaratnam (1983) report on two Denils on the Annaquatucket River in Rhode Island, which successfully passed alewives, and several Denils in Massachusetts and New York State. Undoubtedly other fishways of the weir type have been installed as have fish lifts or elevators on the Holyoke River in Massachusetts for the passage of large numbers of shad.

1.5.2 Western Europe

The weir type of fishway in a variety of configurations has been used in Western Europe since early times. In addition there has been continuing interest in Denil fishways, and much experimental work has been done on this type in the laboratory.

The design most often recommended, however, has been the one based on the work of the Committee on Fish Passes (1942) and variations of it. For high dams (30 m or more), the fish lock has been used.

In Iceland, natural obstructions and low dams have been the main reason for installing fishways, and many of the weir type are installed for the Atlantic salmon and trout there.

In Ireland, fishways at dams have been reviewed by McGrath (1960). He lists a weir type fishway at Parteen Weir on the River Shannon, height 26 ft; submerged orifice fishways at two dams on the River Erne having heights of 94 and 33 ft; Borland fish locks at three dams on the River Liffey and River Lee with heights of 58, 99, and 45 ft; and other variations of fish locks at Ardnacrusha on the River Shannon and on the River Clady in County Donegal. He reports further that as many as 6000 Atlantic salmon have passed through the submerged orifice fishways on the River Erne annually, and lesser numbers at the other installations. The main migratory fish in Ireland are listed as Atlantic salmon, sea trout, and eels, and it is known that all these species except eels utilize the fishways and fish locks noted. The eels present a special problem, and eel passes are included as extra facilities where required. The eel passes normally consist of a narrow trough or pipe lined with synthetic bristles and covered with a flow of water. They will be described in more detail later in this text.

In England and Wales, according to Beach (1984), only the weir type and the Denil type fishways are commonly installed. While other types such as the weir with submerged orifice are not excluded, the concentration seems to be on the weir with notched baffle. In fact, allowable dimensions are specified such as the minimum pool size, minimum depth, size of notches, and even the thickness of baffle. For the Denil, a design based on the Committee on Fish Passes (1942) design is specified, although the dimensions are flexible in order to accommodate the species, size of fish, and conditions at the dam. The conditions of flow at the type of dam constructed in England and Wales must be fairly stable, as no consideration is given in Beach's report to variations in headwater of more than a meter, and very little to tailwater and its effect on entrance conditions beyond saying, "The pass entrance should be located easily by the fish at all flows." The species passed are listed as Atlantic salmon and sea trout. Other species, labeled "coarse fish" are present but "do not cause a significant problem." European eels are present also, but are able to ascend any obstructions occurring, according to Beach.

In Scotland, the submerged orifice with weir type of pass has been used with success for Atlantic salmon, but passes of the weir and overflow type are also used as are fish locks of the Borland type.

In Denmark, according to a report by Lonnebjerg (1980), there are probably more than 1000 dams and weirs preventing salmon from ascending, and only a few of these have fish passes. He describes a Denil fish pass at Arup Molle on the Rohden-Arum River, which was extensively tested at Horsens before installation. It passed brook and rainbow trout down to 10 in. in length in a test, and is considered successful. Another Denil was installed at Tange power station on the Gudena River, surmounting a height of about 10 m. It consisted of eight sections of about 6.5 m each and seven resting pools. The sections were based on the Committee on Fish Passes (1942) design, with single plane baffles and a slope of 1:5.

In Sweden, pool and weir fish passes have been used as well as Denils. The Denil at Hurting reported by Furuskog (1945) and the Alvkarelby and Bergforsen installations are described by McGrath (1955).

In The Netherlands, the country is low and flat and there is no need for fishways to surmount anything higher than 5 m, while most average 1 to 2 m. A simple pool and weir type therefore is used, bypassing the low obstructions. Trout, salmon, and many other species, including cyprinids, are passed in this fashion according to van Haasteren (1987). A special type of pass for eels has been developed, consisting of a vertical pipe with synthetic bristles and a pumped flow of water, which will be described later in Chapter 3, Section 17.

In Germany, there are the same type of installations as in Holland except that farther up the Rhine and its tributaries the water-development projects become higher and call for more conventional pool and weir fishways and higher eel passes.

In France, fishway design and construction has been very active over the last 15 years. Passes of the Denil type, in many configurations, have been tested in the laboratory and constructed in prototype. In addition, passes of the pool and weir and the vertical slot types have been used in the continuing rehabilitation of the Atlantic salmon runs, as well as for sea-run trout and shad. Larinier (1983) has published a guide for technicians faced with migratory fish problems in France, which outlines the data necessary for the design of fish facilities under most conditions and describes the types of fish facilities available to meet these conditions. Fish locks and fish elevators are described as well as the fishways mentioned above.

1.5.3 Eastern Europe And Russia

Zarnecki (1960) reports sea trout as having ascended "suitable pool passes for salmon at two dams for salmon on a tributary of the Vistula River in Poland, one 10 m and the other 32 m high." Another publication, by Sakowicz and Zarnecki (1962), reviewed fish passes in Western and Eastern Europe and concluded that pool and weir passes were satisfactory for salmon, but that other types, such as the Denil, had proven to be unsatisfactory, particularly in Eastern Europe (Russia). Indications by Pavlov (1989) are that the salmonid migration problem is satisfactorily taken care of by standard pool and weir fishways of the type covered later in this text, or by lifts or locks where applicable. However, a major problem is described by Pavlov on the rivers of the south slope of the European region of the former U.S.S.R. This is in the basins of the Caspian, Azov, and Black Seas, and is centered on hydroelectric developments and water intakes for other purposes on the Volga, Don, and Kuban Rivers. Protection of the migrations of sturgeon, Atlantic (Caspian) herring, carp, catfish, and Eurasian perch is a major problem in this region, and the many types of structures that have been devised for this purpose are described and evaluated by Pavlov. Among these are elaborate fish lifts or traps, for the capture of upstream migrants below the major hydroelectric dams, and systems of screening the intakes of water-development projects, some of which will be described later in this text. One important factor that makes this protection unique is provision must be made for transport over barriers and away from intakes of many fish that are only one step removed from their larval stage. Thus, they are dealing with migrants that vary from less than 10 up to 30 mm long, which makes the solution much more difficult.

1.5.4 Latin America

The migratory fish in Latin America are of completely different species than any encountered in the northern hemisphere. There has been little research done on such things as their life histories, migratory habits, and swimming ability. It has therefore been difficult to design satisfactory fish facilities at the hundreds of dams built and proposed on the many rivers of this region.

There are commercially valuable populations of mainly diadramous fish that migrate upstream to spawn in certain areas of the river systems at fixed periods of the year. As the first dams were built, the need for fishways to provide passage for these fish became apparent, and they were included in dams starting early in this century. Lacking the basic information noted above, designs were roughly based on European and North American standards for salmonids. Thus, according to Quiros (1988) more than 50 fishways were constructed in the region over the period from about 1910 to the present, most of these being in Brazil. All of the pool and weir type, with various dimensions that seem to bear no relationship to the size of runs, they have met with varying success.

Fish of the genera *Salminus* and *Prochilodus* seem to be the most important, and to a certain extent they resemble salmon in size (over 30 cm in length) and their swimming and jumping ability. However, through comparison based on what little is known of their swimming ability as stated in Quiros (1988), they appear to be slightly weaker than salmon even though they frequently jump at obstructions. It would seem that standard weir or vertical slot type fishways with drops per baffle of 30 cm or less would satisfactorily pass these fish, provided that the entrance conditions and other criteria outlined later in this text were followed.

In their more recent projects, from 1980 onward, Brazil and Argentina have elected to use Borland type fish locks for dams over about 20 m in height, on the basis that fishways are too expensive to install. These locks have exhibited the usual weakness of not being able to handle large numbers of fish over a short period. In addition the attraction and collection systems described by Quiros are highly inadequate.

The most recently planned installations are of the fish elevator type, based on the premise that passage should be provided for large numbers and all species. These stocks of fish have been more intensively studied, and the fish biomass to be passed has been estimated. The elevators suggested are based on the Russian model described by Kipper and Mileiko (1967). It is thought that this will ensure satisfactory passage of all species. Quiros gives no estimate of installation costs and particularly of maintenance and operational costs, which could be extremely high.

In general, the South American experience seems to be following that of other parts of the world, with rather large variations in success, because of lack of knowledge of the special requirements of the species involved and lack of application of all the criteria needed for good fishway design.

1.5.5 Africa

In Africa, the need for efficient fishways has become apparent only in recent years. The oldest fishways are at Jebel Aulia Dam on the Nile in the Sudan, and some minor stream installations in South Africa. The installation in Jebel Aulia was for the purpose of assisting the migration of Nile perch. Bernacsek (1984) reported it to be functioning unsatisfactorily. Runs of anadromous shad are known to frequent rivers in Morocco and other coastal Mediterranean countries, but to date fishways in the dams in this area have been either overlooked or considered unnecessary.

The vast area of Central Africa, being tropical, has species of fish that grow very fast, and many are readily adaptable to new conditions created by dams and reservoirs. Hence the need for fishways at the many dams in the region has been considered minimal. No doubt with further study, conditions will become apparent for a fishway in a new dam to prove to have real economic benefits, but to date this has not been the case.

In Southern Africa however, where there are many small dams and weirs on the coastal rivers, there is a growing realization of the need for fish passage at many of these dams. The fishways in the dams that do have them have been based on existing European or North American designs for salmonids; thus, they are successful for passing trout but do not meet the needs of other species. Their real needs have been found more recently to be passage for freshwater mullet and other catadromous species that are unable to ascend the fishways designed for salmon and trout. According to Bok (1984), the South Africans are now entering on a research program to define the requirements of these species that will enable them to design satisfactory fish facilities.

1.5.6 Australia and New Zealand

Harris (1984) has described a survey that identified 29 fishways on coastal streams of southeastern Australia. These represented about 9% of all dams, weirs, and causeways in the area. Of the 29 passes, 18 were 2 m or less in height, and the rest ranged up to 8 m. The general conclusion of the survey was that about 75% of these fishways were operating unsatisfactorily. Their design had been based on early designs for salmonid fishways in Europe and North America, and a new orientation to the needs of indigenous species was required. These species, all of which are migratory at some stage of their life cycle, were named as Australian bass, shortfin and New Zealand longfin eels, striped mullet, rainbow fish, shrimp, bullrout, sand mullet, yellow perchlet, rainbow trout, lamprey, and barramundi perch. It was suggested that greater care was necessary in adopting the design principles from North America and Europe, together with new research on the life cycles and swimming abilities of the species under consideration.

1.5.7 China

Xiangke Lu (1986, 1988) notes that 40 fish passes have been constructed at water-control sluice gates in the country. Twenty-nine were constructed in Jiangsu Province, but only three have shown good results. They were constructed mainly to pass Japanese eel, mud crabs, and four major species of carp. In addition, a fishway was constructed at the 15-m hydropower dam at Fuchunjiang in Zhejiang Province but was later abandoned because of poor results.

It should be noted that China has a vast system of reservoirs, numbering more than 85,000. The fisheries of these reservoirs are intensively exploited and are maintained by being stocked mainly with the four major Chinese carps. Because of this policy, little need has been felt for fishways at the outlets to help maintain stocks.

1.5.8 Japan

In Japan, there are almost 1400 fishways, most of which are the pool and weir type in various forms. A small number (about 0.1%) are Denil fishways, and about the same small percentage are fish locks and special eel fishways (Sasanabe, 1990).

The pool and weir fishways are intended mainly to provide passage upstream for the ayu. This anadromous fish, about 25 cm in length, is very valuable to the Japanese people, providing a desirable part of their diet in smoked and other forms. The pool and weir fishways are of varying effectiveness, and there seems to be little rationality in the design of most of them. There is, however, a current effort to improve their design.

1.5.9 Southeast Asia

Pantulu (1988) describes a fish pass in Thailand and one in India. He states that no assessment has been made of the one in Thailand, although it was supposed to benefit some of the cyprinid and siluroid fishes. Pantulu (1984) states that the Mekong River has many migratory fish, some diadromous (threadfin, perch, catfish, and herring) and other main-channel migrants such as the giant catfish and freshwater prawn. These are certain to cause problems of the Mekong is ever dammed.

Pantulu refers to a fish pass at the Farakka Barrage on the Ganga River in Northern India, but is not aware of any assessment of its performance. Because India, Pakistan, and Bangladesh have anadromous species (e.g., Hilsa), these countries undoubtedly will give careful consideration to the possible use of fishways and other fish facilities should the need arise.

Figure 1.4 shows roughly where fishways and fish facilities have been used in the world to date. As suggested earlier, the usual practice in designing fishways at dams is first to make a preliminary choice of fishway type based on general considerations such as water quantity available, probable fluctuations in water level in the head pond and tailwater pool, and *on the record of experience in passing the fish under consideration*. This preliminary choice may be changed at a later date, after all other considerations have been studied in more detail, but it serves as a starting point in the procedure of designing a fishway for a dam.

1.6 A FIRST WORD ABOUT SWIMMING SPEEDS

In designing facilities for migratory fish, in both upstream and downstream directions, the first question that occurs concerns the swimming ability of the migrants. What is their normal cruising speed and their burst speed? These are important questions in the design of a fishway, for example, no

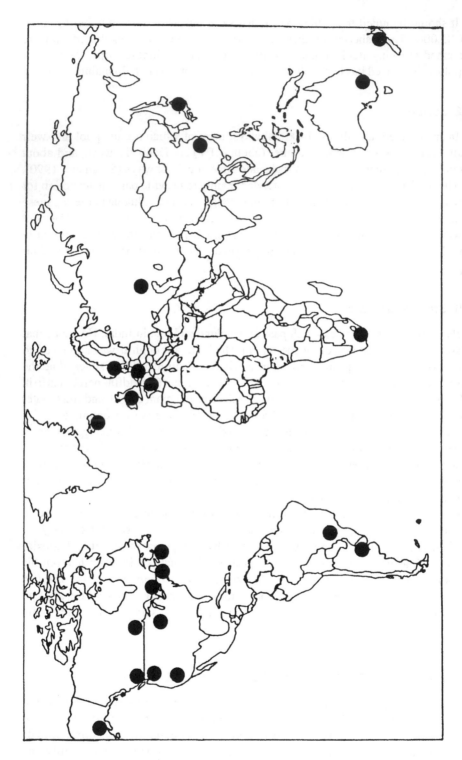

Figure 1.4 Map of the world showing the location of most of the fish facilities described in this chapter.

matter what the type. The velocity of the water over the weirs or through the orifices or slots must be less than the burst speeds, and the velocity in the pools must be less than the cruising speeds. Bell (1984) defines three levels of speeds that are critical in design as follows:

- **Cruising** — A speed that can be maintained for long periods of time (hours)
- **Sustained** — A speed that can be maintained for minutes
- **Darting (or burst)** — A single effort, not sustainable

Beach (1984) defines two of these speeds only, cruising and burst, in terms of utilization by the fish of its aerobic (or red) muscle, and its anaerobic (or white) muscle, respectively. He suggests a gradual transition between these two speeds. The only variables are the temperature of the water and the fish length. He has developed two graphs based primarily on the work of Wardle and Zhou in the U.K., which are shown in Figures 1.5 (a) and (b).

His curves are based on the assumption that all fish of equal length have the same swimming speed, but it must be remembered that the work on which he has based this assumption has been done on cold-water fish and that in his case, it is applied mainly to salmonids. It is doubtful if the warm-water fish could be included in such a simplified graph.

Pavlov (1989), has added two more critical speeds to the foregoing, in dealing with bottom-dwelling and pelagic species of the plains region of southern European Russia. This region includes the basins of Volga, Don, and Kuban Rivers. His two additional velocities, which apply mainly to juvenile fish, are as follows:

- **Threshold velocity** — The minimum current velocity that leads to the appearance of fish orientation against the current (values between 1 and 30 cm/s)
- **Critical velocity** — The minimum current velocity at which fish begin to be carried away by the water flow

It will be noted that these are water velocities rather than swimming speeds, but the critical velocity is at least comparable to the maximum sustained speed, as defined earlier, and the burst speed. Figure 1.6 shows a graph of critical current velocities for various species plotted against length of fish. It will be noted that there is considerable variation between species of the same length. This would seem to indicate that the species of fish and the smaller size range investigated by Pavlov are much more sensitive to velocity changes, a factor that must be taken into account in design of fish facilities, particularly for screening applications and for fish elevators designed to attract very young fish.

All of these considerations are important in designing fish facilities for both upstream and downstream migrants, as will be explained further in the various chapters that follow.

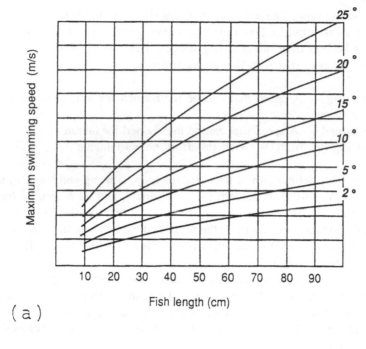

(a)

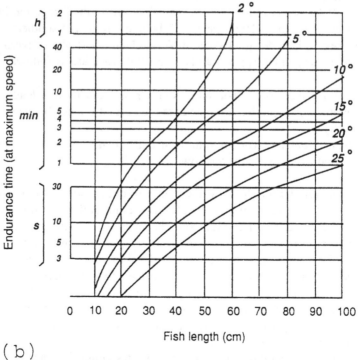

(b)

Figure 1.5 (a) Maximum swimming speeds against fish length for various temperatures; (b) endurance at various speeds against fish length at various temperatures. Temperatures are given in degrees Celsius. (From Beach, M.A., 1984. Fish. Res. Tech. Rep. No. 78, Ministry of Agric. Fish. Food, Lowestoft, England. 46 pp. With permission.)

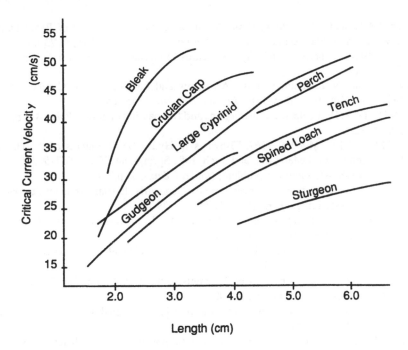

Figure 1.6 Critical current velocities for various species. (After Pavlov, D.S., 1989.)

1.7 LITERATURE CITED

Beach, M.A., 1984. Fish Pass Design, Fish. Res. Tech. Rep. No. 78, Min. Agric. Fish. Food, Lowestoft, England. 46 pp.

Bell, M.C., 1984. Fisheries Handbook of Engineering Requirements and Biological Criteria, U.S. Army Corps of Engineers, North Pac. Div., Portland, OR. 290 pp.

Bernacsek, G.M., 1984. Dam Design and Operation to Optimize Fish Production in Impounded River Basins, CIFA Tech. Pap. 11. 98 pp.

Bok, A.H., 1984. Freshwater mullet in the Eastern Cape — A strong case for fish ladders, *Naturalist,* 28(3).

Cleugh, T.R. and L.R. Russell, 1980. Radio Tracking Chinook Salmon to Determine Migration Delay at Whitehorse Rapids Dam, Fish. Mar. Ser. Manu. Rep. No. 1459. 39 pp.

Committee on Fish Passes, 1942. Report of the Committee on Fish Passes, British Institution of Civil Engineers, William Clowes and Sons, London. 59 pp.

Conrad V. and H. Jansen, 1987. Personal communication.

Decker, T.F., 1967. Fishways in Maine, Maine Dept. Inland Fish and Game, Augusta, ME. 47 pp.

Deelder, C.L. 1958. Modern fishpasses in The Netherlands, *Prog. Fish Cult.,* 20(4), pp. 151-155.

Denil, G. 1909. Les echelles a poissons et leur application aux barrages de Meuse et d'Ourthe, *Annales des Travaux Publics de Belgique.*

Fish, F.F. and M.G. Hanavan, 1948. A Report upon the Grand Coulee Fish Maintenance Project, 1939–1947, U.S. Fish & Wildlife Serv. Spec. Sci. Rep. No. 55. 63 pp.

Fulton, L.A., H.A. Gangmark, and S.H. Bair, 1953. Trial of a Denil-Type Fish Ladder on Pacific Salmon, U.S. Fish & Wildlife Serv. Spec. Sci. Rep. Fish. No. 99. 16 pp.

Furuskog, W., 1945. A new salmon pass, *Sartryck ur Svensk Fiskeri Tedsrift,* 11, pp. 236–239.

van Haasteren, L.M., 1987. Personal communication.

Hamilton, J.A.R. and F.J. Andrew, 1954. An investigation of the effect of Baker Dam on downstream migrant salmon, *Int. Pac. Salmon Fish. Comm. Bull.*, 6, 73 pp.

Harris, J.H., 1984. A survey of fishways in streams of coastal south-eastern Australia, *Aust. Zool.*, 21(3).

Katopodis, C. and N. Rajaratnam, 1983. A Review and Lab Study of Hydraulics of Denil Fishways, Can. Tech. Rep. Fish. Aquatic Sci. No. 1145. 181 pp.

Kipper, S.M., 1959. Hydroelectric constructions and fish passing facilities, *Rybn. Khoz.*, 35(6), pp. 15–22.

Kipper, Z.M. and I.V. Mileiko, 1967. Fishways in Hydro Developments of the U.S.S.R., Clearinghouse for Federal Sci. and Tech. Inf., Springfield, VA. TT67–51266.

Klykov, A.A., 1958. An important problem, *Nanka i Zhizu*, 1, p. 79.

Larinier, M., 1983. Guide pour la conception des dispositifs de franchissement des barrages pour les poissons migreteurs, *Bull. Fr. Piscic.*, 39 pp.

Lonnebjerg, N., 1980. Fishways of the Denil Type, Meddr. Ferskvansfiskerilab Danm, Fisk-gg. Havenders Silkeborg, No. 1. 107 pp.

McGrath, C.J., 1955. A report on a Study Tour of Fisheries Developments in Sweden, Fish. Br., Dept. Lands, Dublin. 27 pp.

McGrath, C.J., 1960. Dams as barriers or deterrents to the migration of fish, 7th Tech. Mtg. I.U.C.N., Brussels, Vol. IV, pp. 81–92.

McLeod, A.M. and P. Nemenyi, 1939–1940. An Investigation of Fishways, Univ. Iowa, Stud. Eng. Bull. No. 24. 63 pp.

Moffett, J.W., 1949. The first four years of King salmon maintenance below Shasta Dam, Calif. Fish Game, 35(2). pp. 77–102.

Nemenyi, P., 1941. An Annotated Bibliography of Fishways, Univ. Iowa, Stud. Eng. Bull. No. 23. 64 pp.

Orsborn, J.F., 1985. Development of New Concepts in Fish-Ladder Design, Bonneville Power Admin. Proj. No. 82–14. Pts. 1–4.

Pantulu, V.R., 1984. Fish of the lower Mekong Basin. In *The Ecology of River Systems*, Dr. W. Junk Publishers, Dordrecht, The Netherlands.

Pantulu, V.R., 1988. Personal communication.

Pavlov, D.S., 1989. Structures Assisting the Migrations of Non-Salmonid Fish: U.S.S.R., FAO Fisheries Tech. Pap. No. 308, Food and Agriculture Organization of the United Nations, Rome. 97 pp.

Quiros, R., 1988. Structures Assisting Migrations of Fish Other Than Salmonids: Latin America, FAO-COPESCAL Tech. Doc. No. 5, Food and Agriculture Organization of the United Nations, Rome. 50 pp.

Sakowicz, S. and S. Zarnecki, 1962. Pool passes — biological aspects in their construction, *Nauk Rolniczych*, 69D, pp. 5–171.

Sasanabe, S. 1990. Fishway of headworks in Japan, Proc. Int. Symp. on Fishways '90, Gifu, Japan.

Schoning, R.N. and D.R. Johnson, 1956. A Measured Delay in the Migration of Adult Chinook Salmon at Bonneville Dam on the Columbia River, Fish Comm. Oregon, Contrib. No. 23. 16 pp.

U.S. Fish & Wildlife Service, 1948. Review Report on the Columbia River and Tributaries, App. P, Fish & Wildlife. U.S. Army Corps of Engineers, North Pac. Div., Portland, OR.

Washburn & Gillis Assoc., Ltd., 1985. Upstream Fish Passage. Can. Elect. Assoc. Res. Rep. No. 157 G. 340 pp.

Xiangke Lu, 1986. A Review on Reservoir Fisheries in China, FAO Fish. Circ. No. 803. Food and Agriculture Organization of the United Nations, Rome.

Xiangke Lu, 1988. Personal communication.

Zarnecki, S., 1960. Recent changes in the spawning habits of sea trout in the Upper Vistula, *J. Cons. Int. Expl. Mer,* 25(3), pp. 326–331.

Zeimer, G.L., 1962. Steeppass Fishway Development, Alaska Dept. Fish and Game Inf. Leafl. No. 12. 27 pp.

2

FISHWAYS
AT NATURAL OBSTRUCTIONS

2.1 INTRODUCTION

The fundamental concept in the approach to the design of a fishway to overcome a natural obstruction is different from that in the case of a dam. This difference in concept can be explained briefly. First, the natural obstruction to migration is in most cases a part of the natural environment of the fish it affects. The population of migrating fish has presumably become adjusted to some extent to this environment. However, if the obstruction each year takes its toll by reason of direct mortality, or physical impairment as a result of delay or damage, any facilities installed that will reduce this mortality or impairment will be beneficial. It is possible then to think in terms of the most economical installation to produce the biggest benefit, even though the result may still be far short of perfection. With a fishway at a dam, however, the primary aim is usually the ultimate one of providing for **no** delay and **no** physical impairment of the fish, since any such delay or impairment, if not part of the natural environment, will in all probability inevitably result in depressing the population.

This difference in fundamental approach when translated into practical terms means simply that the standard of space requirement or size of fishway may often be less for a fishway at a natural obstruction than one at a dam, other factors being equal.

In addition, of course, there are many differences in details between the two cases. Hydraulic conditions at a natural obstruction are not normally subject to any degree of control as they are at a dam. The wide range of flows that occur make it desirable to have regulating devices in the fishway to enable it to operate efficiently. However, natural obstructions are often in locations where it is most difficult to provide for regular operation and maintenance. At dams, on the other hand, power is often available for automatic operation, or an attendant for manual operation, and maintenance crews are at hand most of the time.

This problem has led to the virtual abandonment of the weir type fishway for use at natural obstructions. The weir type of fishway operates efficiently only within a very narrow range of flows. Since the flow in the fishway is controlled by the upstream weir, the fishway itself can operate efficiently only when river levels are within the range producing the desired flows over this weir. In a few cases it may be practicable to provide for regulation of the fishway flow over a wider range of river levels by means of adjustable weir crests, gates, etc., but in most cases it is not.

The search for a solution to this problem, which is almost as vexing at fishways in low dams where no regular attendant is available, was undoubtedly responsible in part for leading Denil to the development of his fishway. More recently, this search led to the development of the vertical slot baffle. The Denil fishway can be designed to operate through a wider range of river levels than the weir type without serious impairment of its efficiency. This range is not unlimited, however, and is in fact only a matter of a very few feet. The vertical slot fishways at Hell's Gate, on the other hand, have operated successfully over periods during which the range in water levels has been as much as 45 ft. Furthermore, there appears to be no reason, other than structural limitations, why there should be any upper limit to this range. The only drawback is that the efficient functioning of the Hell's Gate and other vertical slot fishways depends on the variation in river levels being in phase and reasonably equal above and below the fishway. Fortunately this condition usually exists at natural obstructions. In the rare cases where it does not exist, then even the vertical slot fishway is reduced in efficiency, and it may be necessary to resort to controls of one form or another.

2.2 THE VERTICAL SLOT BAFFLE

Because of its success in application to the delicately balanced requirements of Pacific salmon, the vertical slot baffle will be described in detail, and an indication will be given of its various applications to date. It also appears that the qualities shown in its use with Pacific salmon in all probability will be equally advantageous in application to other species of fish.

The baffle evolved by the International Pacific Salmon Fisheries Commission for the main fishways at Hell's Gate is shown in Figure 2.1. The flow pattern in each pool is shown diagrammatically by the arrows in Figure 2.2. It will be noted that the jets from the two slots meet in the center of the fishway just upstream of the next baffle downstream. The fact that these jets meet at an angle to each other improves the energy dissipation in each pool. If they were parallel the cushioning area of the pool would have to be relied on entirely for energy dissipation. The small wings projecting upstream from each center baffle also assist by directing part of the jet back upstream, so that there is a small component acting in the direction opposite to the main jet through each baffle. The small wing walls projecting from the fishway walls immediately downstream from the center baffle wall help to direct the main jets toward the center of the fishway.

The baffle dimensions shown were arrived at after thorough model testing at the University of Washington in Seattle, by C.W. Harris, Professor of Hydraulics, University of Washington, Seattle and E.S. Pretious, Professor, Department of Civil Engineering, University of British Columbia (unpublished work). Many combinations of baffle shapes, placing, slot widths, etc. were tested, and the one producing the pool with the least turbulence and the fewest other undesirable hydraulic characteristics was finally selected.

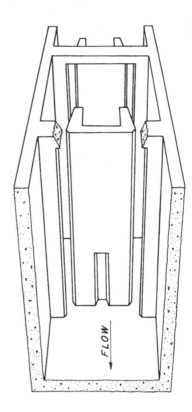

**Figure 2.1 The Hell's Gate
paired vertical slot fishway
baffle.**

Some of the advantages of the vertical slot design with relation to behavior of
adult salmon are as follows:

1. Ascent of the fishway is possible at any depth the fish chooses. There can
   be considerable variation in depth selection by the fish, depending on the
   time of day, light conditions in the fishway, turbidity of water, etc.
2. The path of a fish ascending the fishway is not tortuous. Many fishway
   designers in recent years have expressed the view that fishways with
   staggered orifices or slots in the weirs or baffles are not entirely satisfactory. Apparently this view has resulted from observations of fish, which
   have shown difficulty or distress in passing through or over a series of such
   baffles. The reason may be that the fish orient themselves to a certain
   extent in relation to their distance from the fishway walls. The fact that
   passage through the slots of the vertical slot fishway is near the walls, in
   addition to the symmetry of the baffles, may have some bearing on their
   success.
3. Conditions for resting in the pools are satisfactory, if required. The physiological requirements in this respect are not clearly understood, but recent
   tests have indicated that fish tend to accumulate in long fishways, so that
   it is still generally believed that areas in which fish can maintain position
   without undue stress (if this can be described as resting) are a necessary
   requirement.

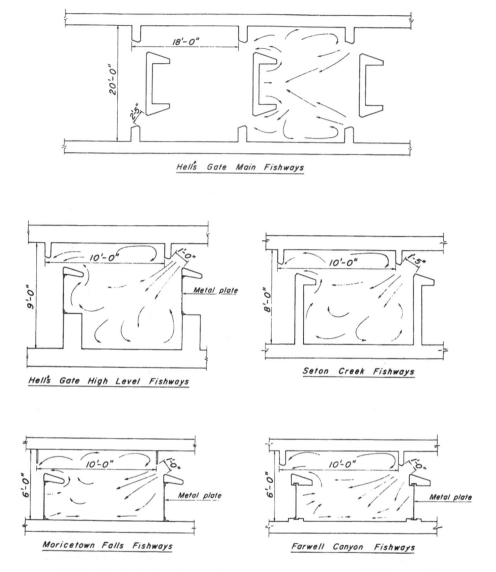

**Figure 2.2   Baffle plans and flow patterns for the main fishways at Hell's Gate
and for several smaller fishways in Western Canada.**

It may be of interest here to review the design criteria and assumptions used for the main
fishways at Hell's Gate in arriving at the pool dimensions of 20 ft wide, 18 ft long, and 6 ft
deep at minimum river stage for fish passage. These are discussed by Jackson (1950). From
a review of tagging data for Hell's Gate and an estimate of the future migration pattern in the
Fraser River, he arrived at a maximum rate of migration of 500 fish per minute, or 20,000 fish
per hour in the peak hour. He then assumed that a volume of 2 ft³ of water was the minimum
requirement for each fish, and finally estimated a time of migration through the fishway and
through each pool. This was set at 5 min per pool, or 45 min for the fishway, which he felt
was conservative.

By combining these criteria, he showed that each pool at minimum river stage would accommodate 1080 salmon. With a rate of migration of 500 fish per minute or 250 per minute along each bank of the river, and 5 min per pool, 1250 fish would be the maximum number a pool would be required to accommodate. This is slightly larger than the pool capacity available, but on an hourly basis, the capacity proves to be adequate.

This is the first known recorded instance where a rationalization of fishway dimensions was attempted by using the volume required to accommodate peak runs. Up until the time Jackson's analysis was published, the maximum figures he used had not been approached in practice, so that the ultimate limit of the fishway capacity had not been tested. Tagging results had shown that they were satisfactory for the numbers using them.

2.3 APPLICATIONS OF THE VERTICAL SLOT BAFFLE TO SMALLER FISHWAYS

The vertical slot baffle used on the main fishways at Hell's Gate required extensive laboratory tests, as pointed out previously. Subsequent to the construction of the Hell's Gate fishways, however, a need arose for a smaller fishway on tributaries of the Fraser River. The natural obstruction at Farwell Canyon on the Chilcotin River was shown by tagging to delay the Chilko River salmon runs. This canyon was remote and virtually inaccessible in winter so that the automatic adjusting hydraulic features of the baffle used at Hell's Gate were highly desirable. However, the maximum expected hourly peak run was only a small fraction of that at Hell's Gate, so that a smaller, more economical size of fishway was needed. For this purpose, the baffle plan used at Hell's Gate was arbitrarily halved along the centerline, and in addition the scale was reduced to approximately one half. The Farwell Canyon fishways were designed using this baffle plan, with extensive laboratory testing of the baffle before the fishways were constructed. The baffle appeared to operate reasonably satisfactorily, and has since been further tested and improved. The original Farwell Canyon plan is shown in Figure 2.2, along with other, more recent fishway designs, all of which have been thoroughly tested in the laboratory and proved in field operation.

It will be noted that this smaller baffle is not symmetrical about the centerline, and perhaps for this reason it appears to be more sensitive to minor changes in dimensions of the component parts. Because of the loss of symmetry, the advantage in energy dissipation resulting from the mixing of the jets from paired slots is lost. Practically all the energy dissipation has to be accomplished by mixing in the cushioning area of the pool, and for this reason the direction of the single jet is very important. There is a tendency for the jet to turn directly downstream toward the next slot opening, which results in a velocity carryover from pool to pool. This has been overcome by adjusting the dimensions of the baffle components, and by addition of a sill of up to 12 in. in height across the slot on the fishway floor, as shown in Figure 2.3. The sill is more effective for fishways at low depths and tends to decrease in efficiency as the fishway depth of flow increases.

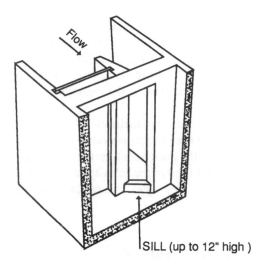

Figure 2.3 The Farwell Canyon single vertical slot fishway baffle, showing the floor sill at the bottom of the slot to improve flow conditions in the pools.

SILL (up to 12" high)

The laboratory testing of all the various baffle plans described, and others as well, has been done by agencies such as the International Pacific Salmon Fisheries Commission and the Department of Fisheries of Canada. Unfortunately most of the results were not published, but a general comment will indicate the type of data collected, which is on file with these agencies.

The Farwell Canyon baffle plan was tested in hydraulic models by Pretious and Andrew (1948) in pool sizes (prototype) of 8 by 6 ft, of 10 by 6 ft, and of 8 by 8 ft, the first dimension being length and the second width in all cases. With a head on each baffle varying from 0.58 ft to 1.30 ft and a slot width of 12 in., the coefficient of discharge (C_d) was found to vary between 0.67 and 0.84, and the jet velocity to vary between 5.16 and 10.3 ft/sec. Varying sill heights were also tested in combination with various pool sizes, and qualitative observations of turbulence, surge, upwellings and draw-downs, aeration, etc., were meticulously recorded along with the aforementioned quantitative observations.

In general, the prototype vertical slot fishways that have to date been constructed to the smaller plans have performed satisfactorily and exhibited the same three advantages noted for the Hell's Gate plan, namely, (1) ascent by the fish is possible at any depth, (2) the upstream path for ascending fish is not tortuous, and (3) conditions for resting en route are satisfactory.

A still smaller fishway was designed by the author for the Fish and Game Branch of the British Columbia Department of Recreation and Conservation. It is intended for use in trout streams where the fish are considerably smaller than the mature salmon. The Farwell Canyon dimensions were reduced by almost one half, resulting in a pool 4 ft wide by 5 ft long with a vertical slot 7½ in. wide. The baffles were constructed of wood, as shown in Figure 2.4, with walls of either concrete, as shown, or timber. Installations of this size have apparently operated satisfactorily to date, although it should be kept in mind that they are definitely limited in the size of fish they are capable of passing. The following general conclusions can be made regarding the smaller vertical slot baffles developed from the Farwell Canyon plan:

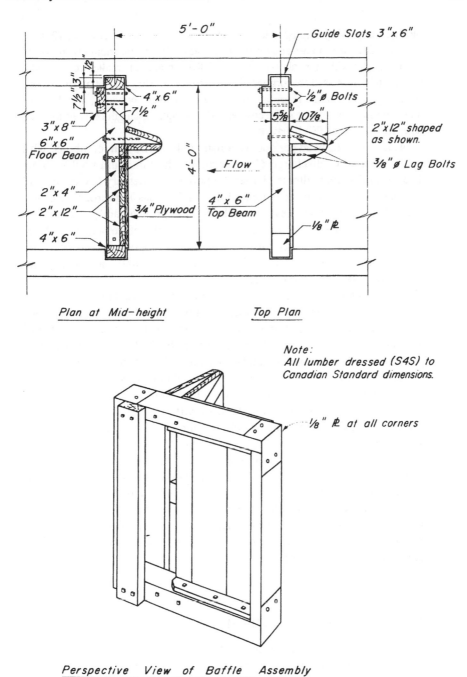

Plan at Mid-height   Top Plan

Note:
All lumber dressed (S4S) to
Canadian Standard dimensions.

⅛" ℞ at all corners

Perspective   View   of   Baffle   Assembly

Figure 2.4   Baffle plan of small fishway for trout using a single vertical slot. This design uses a timber baffle and concrete fishway walls.

1. **Slot width** — Very limited field experience with slot widths less than 12 in. in fishways designed to pass adult salmon, suggests that 12 in. should be regarded as a minimum width for salmon weighing 5 lb or more. For smaller trout, the slot width can be decreased proportionately, with a probable lower limit of 6 in. for fish weighing 2 lb or less. There is no indication to date, however, whether this limitation in width of slot is directly related to weight or length only, or to some other factor such as fish behavior.

2. **Pool size** — If the head per pool is constant, the slot width (combined with the coefficient of discharge C_d) determines the minimum pool volume. The larger the slot width, the larger the pool size necessary to cushion and dissipate the energy of the increased rate of flow passing through the slot. For a 12-in. slot, a pool measuring 6 by 8 ft has proved barely satisfactory. A 6- by 10-ft pool shows less turbulence and is therefore better, while an 8- by 10-ft pool is the least turbulent and most satisfactory for the 12-in. slot. Similarly for 6-in. slots, 3- by 4-ft pools are barely satisfactory and 4- by 5-ft pools are much better.

3. **Head per baffle** — The head loss per pool may be varied to decrease flow and therefore turbulence, to compensate for pools of smaller dimensions where they are required. For a 6- by 8-ft pool, the head per baffle should generally be kept less than 1 ft, but for the 6- by 10-ft and the 8- by 10-ft pools, 1 ft of head provides satisfactory control of turbulence. Reduction of turbulence may not be the only criterion, of course. Where 1 ft of head per baffle has been found satisfactory for the more agile species of Pacific salmon and steelhead, something less, usually of the order of 0.75 ft, has been found desirable for pink and chum salmon, which are weaker swimmers. This requirement is of course related to the fish's swimming ability in the high velocities prevailing in the slots. For other species the swimming ability and behavior of the fish must be taken into account. The burst speed of the fish must not be exceeded in the slots, and that is why it is so important to know what this speed is for the species in question. In addition any behavioral characteristics, such as the preference of American shad to travel through a fishway in groups, can be important as they can rule out a narrow slot.

2.4 DENIL FISHWAYS FOR NATURAL OBSTRUCTIONS

The Denil fishway has some advantages that make it suitable for limited use at natural obstructions. In common with the vertical slot fishway, it allows the fish some choice of swimming depth and have a route of ascent that is not tortuous. It lacks resting area within the fishway, but this can be provided by connecting a series of resting pools of adequate volume by short lengths of fishway. Because of the excellent energy dissipation, the Denil is able to pass a larger volume of water than other passes of comparable cross section, resulting in improved attraction of fish at the entrance. This larger volume is a definite advantage at a natural obstruction where ample flow is available and fish might have difficulty in finding the entrance to the fishway. In addition, the shorter length of the Denil pass and its steeper gradient (1:3 to 1:5 compared to 1:5 to 1:10 for a small vertical slot) mean maximum economy in construction. Some of this is lost in longer fishways, however, because of the need to provide resting pools at intervals.

The disadvantages of the Denil are that it has a limited range of flows because of its limited accommodation to changes in depth and its intricate baffle design, which results in problems in both construction and maintenance.

The Denil fishway has a long history of development, with much laboratory work accompanying it. Reference to Figure 2.5 will illustrate this to some extent. Figure 2.5(a) shows the model of the original Denil fish pass as developed by Denil in 1908. While this was the most sophisticated design and best energy dissipator to date, it was recognized by Denil and others that it could not be constructed readily in the field and much work was done on trying to find a simplified structure that would provide the same energy dissipation. Studies were undertaken by McLeod and Nemenyi (1939–1940) and by the Committee on Fish Passes (1942), and the latter evolved a design that has become virtually a standard for the ensuing years; it is shown in Figure 2.5(b). Since the baffles are in one plane, it is readily constructed, and although dimensions and slopes have been varied, it still forms the basis for many of the fishways used today.

Figure 2.5(c) shows a variation of this design, called the *Steep pass,* which was developed in Alaska by Zeimer (1962). It is designed to be constructed out of aluminum sheets, so that it can be prefabricated and flown into remote sites in Alaska. This and other versions of the Committee on Fish Passes design have been thoroughly tested in the laboratory by Katopodis and Rajaratnam (1983). In these tests the slope was varied from 1:10 to 1:31.5, and flows were varied from 2 to 4 cfs (67 to 113 l/s), thus giving depths within a narrow range of approximately 25 to 50 cm. The testing was detailed and thorough and gives a good picture of what to expect from the Denil under comparable conditions in the field.

Other tests have been made on Denils, particularly in Europe, and the literature is quite extensive, but the purpose has been to test them for use at dams, so they will be dealt with in a later chapter.

To summarize the situation with respect to fishways at natural obstructions, it is suggested that first of all, one should observe and record the fluctuations in flow as represented by the variations in water level over the period that fish are migrating or are expected to migrate. If the variation is low, a pool and weir or Denil fishway might be considered, but if high, a vertical slot fishway is recommended. This is summarized as follows:

Range of depths due to variations in flow	Type of fishway recommended at natural obstructions
Relatively constant	Pool and weir or Denil
Up to 5–6 ft, or 2 m	Denil or vertical slot
Over 6 ft, or 2 m	Vertical slot

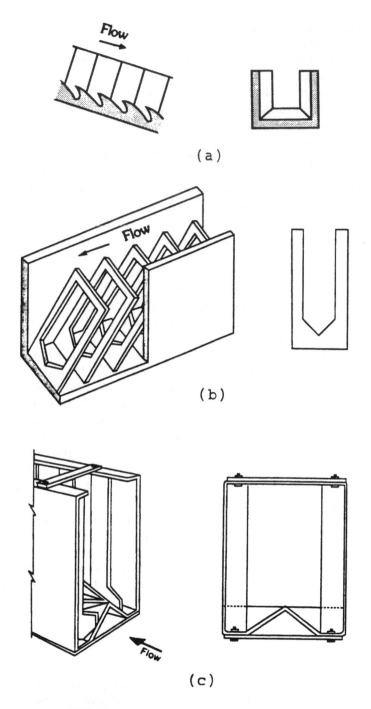

Figure 2.5   Typical Denil fishways. (a) Longitudinal and cross sections of original developed by Denil in 1908; (b) perspective view and cross
section of version recommended by Committee on Fish Passes in
1942; (c) perspective and cross section of steep pass version developed by Zeimer in 1962.

2.5 A PROCEDURE FOR INSTALLATION OF FISHWAYS AT NATURAL OBSTRUCTIONS

The following procedure for the installation of fishways for migratory fish over natural obstructions was developed in British Columbia in the 1940s. While it was developed at Hell's Gate to meet the need to provide passage for the large runs of sockeye salmon migrating up the Fraser River, it has resulted in fishways that satisfactorily pass most other species of Pacific salmon and trout. There is no reason to believe that, with modifications to the part of the procedure dealing with the selection of functional dimensions, it could not be applied to any species of migratory fish at any natural obstruction. Furthermore, while the procedure as developed and outlined here applies to vertical slot fishways, modifications to the section discussing functional dimensions would make the procedure equally applicable to any presently known type of fishway.

2.6 DEFINING THE PROBLEM — BIOLOGICAL DATA

It is fortunate, perhaps, that the large numbers and high value of the salmon obstructed at Hell's Gate warranted an intensive biological investigation before design of the fishways was commenced. This biological investigation has been reported in detail elsewhere, and even though some controversy arose with respect to the scientific evidence resulting, it was abundantly clear from this evidence as well as from observation that a serious obstruction existed at certain water levels (or river stages).

From the point of view of the engineers designing the fishways, the fortunate thing was that the river levels over which the obstruction occurred were quite clearly defined by the biologists. Even if the basis for this definition had been greatly inaccurate, which it was not, the application of a conservative safety factor would still have resulted in a narrowing of the design limits from those that would have had to be imposed had there been no such biological data. It has been the writer's experience that any biological data, even though not exact or complete, is of value to the engineer in arriving at the best design of fishways.

Following Hell's Gate, the many fishways that have been constructed at natural obstructions were designed using generally less biological data, but in all cases some such data have been available. A tagging program to indicate the severity of the obstruction will normally yield such other information as the rate of migration in the river and the timing of the start, peak, and end of migration. It may also indicate the maximum numbers of fish migrating per unit of time. Even without a tagging program, which may not be justified in cases where the runs are small, observation by trained personnel can very often establish the timing of the start, peak, and end of the run, together with some idea of the numbers of fish to be passed per unit of time. These data are necessary prerequisites to the design of the most efficient and economical fish-passage facilities.

2.7 ENGINEERING SURVEYS AND FIELDWORK

The first step is to make a thorough topographic survey of the area of the obstruction, including, if possible, the contours of the river bottom. For obtaining the

**Figure 2.6 Surveying a natural obstruction. The instrument is set up over a
permanent hub or marker consisting of an iron pin set in bedrock.**

contours, and as a basis for all surveys, a primary system of permanent *hubs* or survey
markers should be established in the area, as shown in Figure 2.6. If the area is small,
two of these might be sufficient, with an accurately measured baseline between. For
larger areas, four, arranged in the shape of a quadrilateral, with two on each bank, or
more if necessary, can be used. These markers should be as permanent as possible,
so that they will not become lost during the period before construction starts or even
during construction. An iron pin cemented into a hole drilled in bedrock is excellent
for this purpose. If bedrock is not available, a long iron pin driven in the ground or
a durable wooden stake might suffice. These control hubs must then be located,
preferably to an accuracy of 1 in 10,000 or better. One or more closed level circuits
will establish the elevations of the markers to any desired accuracy. Contours can be
obtained by working from the primary system of hubs by any acceptable survey
method such as cross sectioning, plane table, or random stadia sights or shots
recorded in a field survey book for later plotting.

Natural obstructions usually occur at points where the topography is extremely
irregular, often including bedrock with vertical faces and even caves, etc., which are

difficult to map accurately. The method used is largely a matter of choice to suit the local conditions, and any of the methods discussed, or others, might be found suitable under different conditions.

Contours of the river bottom present a special problem that can be quite difficult to solve unless river flows are low enough to permit wading at some time during the year. If flows are not low enough for wading, and it is not possible to use a boat, some success might be possible using a cable car. It might not be necessary in all cases to obtain bottom contours, however, particularly to a high degree of accuracy, but this will have to be left to the judgment of the engineer. If a hydraulic model of the obstruction is to be constructed to scale, fairly accurate bottom contours will be needed to get the best results from the model.

The hydraulic conditions at any obstruction are determined by the topography and the velocity and nature of the flows entering and passing through the reach. The topography alone does not give the full picture, therefore, and additional data on the hydraulic conditions are required. These conditions are usually complex, and cannot be calculated or rationalized from limited data. Methods had to be developed to obtain data that would give as complete a picture as possible of the conditions affecting the ascent of fish.

Since the high velocity of water, and to a lesser extent turbulence and aeration, are the prime factors preventing ascent of fish, a method of recording comparative magnitude, direction, and location of velocities was developed. Recognizing that it would not be practical to measure the velocities directly by current meter, another, indirect method had to be found to determine at least the comparative velocities along the marginal areas of the river up which the fish pass. The marginal water-surface profile was used to fill this need.

The velocity in an open channel is a function of the slope of the water surface if the flow is steady. A profile of the water surface taken along the bank of a river provides a record of the slope at any point. This, in conjunction with the plan, gives an indication of the comparative magnitude, direction, and location of the velocities. As velocities increase, the migrating fish are forced into the marginal areas of the stream, where they are able to utilize the lower velocities and back eddies to further their progress upstream. The water-surface profile, when taken immediately adjacent to this path, provides a valuable record of the velocity conditions met by the fish at any particular position. If profiles are taken at a number of different river stages, a complete picture of the comparative marginal velocities can be produced.

The range of river discharges observed should include the full range likely to occur during migration. From these data, a picture of the conditions to be encountered by the fish under all conditions likely to occur is readily available.

In the writer's experience, the best method found to date for obtaining the water-surface profiles is by random stadia sights or shots taken from the permanent control hubs in the same manner the topography was obtained. Because records from several years might be desirable before commencing a project, the permanent survey markers recommended earlier are even more valuable for the purpose of obtaining water-surface profiles both before and after construction.

The standard of quality of the water-surface profiles obtained will depend on the ingenuity and ability of the survey team. It is necessary, for example, to ensure that all important changes in the slope of the profile are recorded, without sacrificing efficiency by taking too many measurements. It is also necessary to judge the midpoint of the surge (where the surge is significant) for each measurement, and some experience is required on the part of the rodman in order to do this effectively. In some cases an important point on the profile may be at the bottom of a sheer cliff or at some other inaccessible point, and considerable ingenuity may be necessary to obtain the required water-surface elevation here. One method that has been used in such cases is to locate by stadia a point at the top of the cliff and measure the vertical distance down to the water surface by a tape stretched tight by a small weight attached to its lower end.

Other hydraulic data should also be recorded during the survey, such as the location of points of turbulence and upwelling, and the intensity and location of points of much surge. These data can then be related to biological observations of the behavior of fish in the marginal areas, in order to ensure that the fishway will be constructed in the best location.

Perhaps the most important hydraulic data to be gathered are river discharge measurements. As pointed out in the section of the Appendix A that deals with measurement of stream flow, stream flow data will in many cases be available from an official source such as a government water resource agency. If this is the case, and the record is a continuing one, the local hydraulic data taken by the fisheries worker can be readily correlated with the discharges in that river for the same dates. In this way it is possible to determine what the conditions were that the fish faced at the obstruction over the years covered by the discharge records. From river discharge data it is possible to plot a composite hydrograph, as shown in Figure 2.7, which forms a convenient picture of the discharge pattern of the river throughout the year.

If discharge data are not available from previous years, it might be necessary for the engineer to use other methods. While the actual river flow in this case may not be too important, it is essential in the design of the fishway that there be some data on the range of river **levels** to be expected during the migration period. Setting one or more water level gauges above and below the obstruction and ensuring that they are read periodically through at least one year's migration period is the minimum acceptable procedure. This should be done in any case even when official river discharge measurements are available. The gauge readings can then be used to determine the relationship between water levels at the obstruction and discharges at the official gauging station, which may be some distance away. If the gauges can be read over a period of more than one year, so much the better. Lacking discharge measurements, these gauge readings can then be used to determine the minimum and maximum limits of operation of the fishway, without the necessity of metering the stream to determine the corresponding flows.

Other engineering field work is required for the design and construction of the fishway itself. Foundation conditions must be known if a satisfactory hydraulic structure is to be designed. This will require some degree of observation and examination by an engineer with experience in this work or by a geologist. For small fishways, a surface examination by an experienced engineer may be sufficient, while for larger structures involving greater capital investments, or in special cases, it may

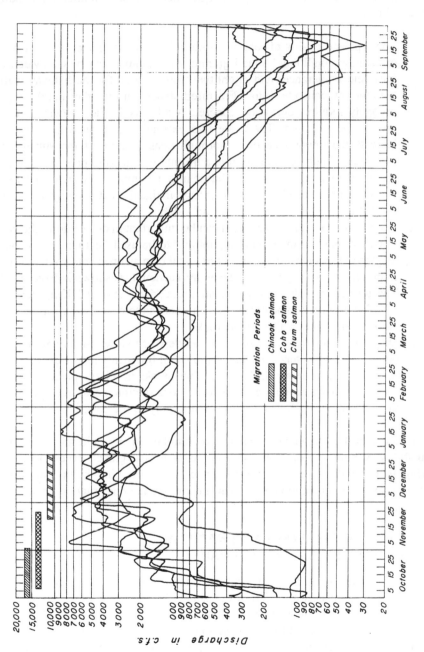

Figure 2.7 Composite hydrograph of a typical river on the Pacific Coast of Canada. It covers a period of six years.

be desirable to have the advice of a competent geologist. In a few cases it may be necessary to supplement surface examinations with subsurface exploratory drilling.

Some of the reasons for giving thorough consideration to foundation conditions are as follows:

1. The type and quality of rock will determine what to expect in the way of excavation conditions. Sound granite, for example, can be excavated to fairly close tolerances and might therefore be used for the fishway walls without any concrete lining. Fractured shales, on the other hand, would probably require reinforced concrete walls inside the fishway cut, to protect the structure. These are two extreme cases, but they serve to indicate the importance of a thorough consideration of the subsurface conditions. Such consideration is a normal engineering procedure in the design of hydraulic structures, but it may be emphasized here that fishways require a consideration as thorough in this respect as many dams and river structures of far larger size and value.
2. The type of rock to be excavated can be important in determining the cost of excavation and the type of equipment to be used. In some cases short lengths of tunnel may be more economical than an open cut, when the cost of protective cover gratings is taken into account, as shown in Figure 2.8. The amount of explosives and the number and length of drill holes per unit of volume of rock to be excavated, and hence the cost of excavation, will all depend on the quality of the rock.
3. Disposal of the excavated material may depend to a certain extent on rock quality. Rock that tends to shatter into small fragments can be disposed of in the river with less risk than rock that breaks into large pieces. This is important to the economics of fishway construction, because disposal on the bank above high water adds considerably to the cost.

Other data that the engineer should accumulate in the field are notes on the availability of construction materials such as sand and gravel; access to the site, having in mind the type of equipment and materials it might be necessary to transport; and accommodation at or near the site for the construction crew.

2.8 DESIGN OF FISHWAYS FOR NATURAL OBSTRUCTIONS — PRELIMINARY

The first stage of the design in most cases is to produce preliminary sketch plans and a preliminary cost estimate. These are usually based on experience alone, but are often the crucial part of the entire procedure, since detailed design is likely to follow the initial layouts closely unless some serious objections develop. The experienced engineer will leave sufficient flexibility in these preliminary plans and estimates to permit any changes that might be necessary at a later date without altering the overall cost. While final, detailed cost estimates can be made at a later date, taking into account the unit quantities of material to be used and all other necessary factors, preliminary estimates can often be based on more general experience.

Figure 2.8 A fishway with a tunnel entrance *(arrow)* **in Western Canada.**

Preliminary estimates prepared from experience or from more detailed considerations, if desired, are valuable in making an assessment of the economic feasibility of a project. The benefits to be derived in increased production by the fishery must be weighed by the fisheries administrator against the estimated cost. Preliminary plans are often useful in giving a clearer picture of the type of project envisaged, and may also be useful in obtaining legal clearances for construction from various agencies and individuals whose property rights might be involved.

If the project appears to be economically justified and is authorized, the next step is to prepare detailed plans. The topographic data obtained in the field must be plotted, and the hydraulic data plotted and analyzed.

A topographic map of the area can be plotted to any convenient scale. Thorough photographic coverage of the area being mapped is always valuable in plotting the contours. Photographs of the river at various stages are also helpful in interpreting the profiles and in choosing the location of the fishway, particularly the entrance.

It may occur to the reader to question at this point why aerial photography and mapping have not been mentioned. There have been rapid advances in aerial photogrammetry over the years, and the usefulness of this tool is acknowledged in other fields of fisheries management, such as the mapping of spawning areas, and the locating of obstructions and rapids in inaccessible river reaches. However, it has been tried and found unsatisfactory for the topographic mapping of obstructions. Underwater contours either are impossible to obtain or at best are of uncertain accuracy. In addition, where rugged, rocky terrain is encountered as is normally the case at an obstruction, the contours obtained even from low-level photography are not accurate enough for the design and construction of fishways. Aerial photography and mapping have been used to advantage, however, to obtain contours of the larger areas surrounding an obstruction, to provide data necessary in laying out access roads, campsites, and disposal areas.

Plotting of the water-surface profiles in the plan and elevation at the same scale as the topographic map is the next step. These can then be superimposed on the base map to determine the shoreline and the water-surface elevation at any desired point for any river stage. All the water-surface profiles should then be projected onto a common baseline and plotted on a separate sheet for each river bank. While these projected profiles tend to exaggerate water-surface slopes, they are very useful in the design of the fishway and for reference purposes later. A typical series of projected water-surface profiles is shown in Figure 2.9.

If river discharge records covering a period of years are available, and a composite hydrograph has been prepared, it is possible to plot the migration period on the hydrograph, as shown in Figure 2.7. It is then a simple matter to determine the range of discharges likely to occur during migration, and the water-surface profiles corresponding to this range can then be used with confidence in the design of the fishways. If the record of river discharges is lacking, then a plot of the daily gauge readings can be substituted for the composite hydrograph. In this case, because a shorter period of record is likely, it is well to allow a liberal safety factor in determining the range of river levels likely to occur during migration and the resulting range of water-surface profiles to use in design.

It should be noted here that the water-surface profiles for each river bank may differ considerably; this is the reason for separating the projected profiles for each bank. A fishway on either bank should be designed to integrate with the profiles occurring along that particular bank.

The first step in finally locating the fishway is to examine the topographic map in conjunction with the projected water-surface profiles. The entrance to the fishway is tentatively located on the profile at the bottom of the steepest drop in the water surface, where velocities are generally the highest (as shown in Figure 2.9) and the corresponding point noted on the topographic map. This point should be checked against biological field data, since it should be the point where fish are actually blocked in their upstream migration by high velocities. The fishway will extend from this point to a safe distance above the obstruction, preferably ending in relatively calm water with low velocity, so the fish will not be swept downstream after passing through the fishway. In some cases, the location of the fish exit may be determined by the length of fishway necessary to accommodate the rise in profile through the obstruction.

There is one further step that might be taken to ensure the best possible location of the fishway to suit the particular physical conditions present at the obstruction, and that is to build a hydraulic model to scale. Scale models of rivers and hydraulic

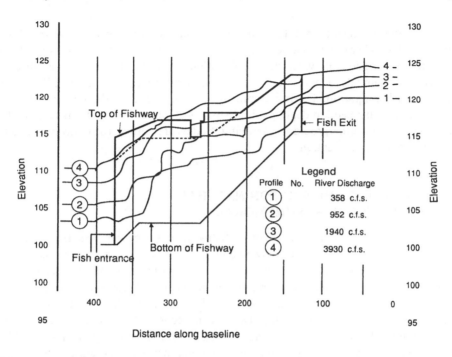

Figure 2.9 **Water-surface profiles on one bank of a river. A fishway was installed to the elevations shown.**

structures have become increasingly recognized in recent years as being extremely valuable aids in the design of river-training and hydraulic structures. They are valuable for two reasons: they ensure the design of more efficient structures and remedial works, and they often result in savings far beyond the cost of the model. Of the two types of hydraulic models, the movable bed and fixed bed, the latter is the one that can yield the most accurate quantitative results and furnish the most clearcut answers to specific hydraulic problems. Because most obstructions occur in rock canyons or at waterfalls where very little bedload accumulates, a fixed-bed model with an undistorted scale is the one that would normally be used to aid in the design of a fishway.

In deciding whether a model is to be used, cost may be a determining factor, as many fishway projects are so small that model construction is not justified. Where one is justified, many advantages are evident. The entrance to the fishway can be varied in the model until the optimum conditions are reached for all required river discharges. In addition, the exit can be better located, and the functional features of the fishway itself can be tested for a wide range of flows. Structural design features can be checked, such as the maximum head to be expected on parts of the structure under varying discharges, and construction problems, such as the conditions to be expected at cofferdams, can also be checked.

The hydraulic model must be carefully constructed, calibrated, and verified by comparison with prototype water-surface profiles, velocities, and flow patterns at comparable discharges. The selection of scales depends mainly on the capacity of the laboratory because in general the larger the model, that is, the closer it is to prototype size, the better the results. A 1:10 scale model is far easier to work with than a 1:25 scale or 1:50 scale model. One reason is, of course, that in models such as the 1:50 scale, minor variations in water-surface elevation are so small that they are difficult to measure accurately, whereas these small differences may be quite significant in terms of fish migration. Further comments on the use of hydraulic models are contained in a later section dealing with fishways at dams. The reader is referred to these sections since the problems involving models are similar for fishways at both dams and natural obstructions.

2.9 DESIGN OF FISHWAYS FOR NATURAL OBSTRUCTIONS — FUNCTIONAL

As a first step in the final design of the fishways, it is necessary to select the gross dimensions of the structure. This may be only a matter of verifying those used in the preliminary estimate, but in any case there is enough background of research and experience to make an attempt at a rational analysis. The object of the analysis is to determine the minimum size and cost of a fishway that will pass the expected maximum run with the least delay. Delay can occur in two major areas: at the entrance to the fishway, and in passage through it. Entrance delay has presumably been minimized by the best positioning of the entrance with the aid of the water-surface profiles as described earlier. Delay in passage is minimized by providing ample volume to accommodate the peak of the run without overcrowding.

The problem then is to find the minimum fishway or pool volume that will meet requirements at minimum river stage. One of the chief unknowns is the volume of water required by each fish. Jackson assumed 2 ft^3 is required for each sockeye salmon under the conditions existing at Hell's Gate. Other criteria have been used elsewhere, where different species of fish of a different size have been encountered. Bell (1984) recommends 0.2 ft^3/lb (4 l/kg), which for the average sockeye at 9 lb is 1.8 ft^3, which is nearly the same as Jackson's recommendation. Bell's more general criterion is recommended for other species. It should be kept in mind that this space requirement applies only to fishways and not to holding and trapping facilities, which are dealt with in another chapter.

It has been found that unless the peak runs to be accommodated are high, the minimum size of fishway (governed by criteria other than the foregoing) is very often adequate. This minimum size is considered to be pools 8 ft wide by 10 ft long, with 2 ft of depth at minimum river stage during migration for salmon or trout in the range of 4 lb or larger. Under some circumstances, pools 7 ft wide by 8 ft long by 2 ft deep at minimum river stage during migration might be permissible. And, as stated earlier, smaller-sized fishways can be employed for smaller trout or smaller fish of other species having behavior patterns and physical capabilities similar to trout.

It might be of interest here to note that Denil (1909) listed among the properties of good fish passes the requirement that "locally restricted cross-sections should have a minimum breadth of 30 cm." This is of interest in that he was probably the first person to recognize that it was necessary to place a minimum limit on the size of restriction, and also because of its remarkable similarity to the minimum suggested by later experience with Pacific salmon arrived at quite independently. The minimum slot width of 12 in. (30.5 cm) recommended earlier conforms very closely to Denil's minimum. As described earlier, this slot width determines the allowable pool size required to dissipate the energy of the jet through the slot from which the minimum pool dimensions described above have been derived.

As noted earlier, in addition to the volume of water required by each fish, an important extra factor is the rate of ascent of a fish through the fishway. It can readily be seen that if a fish ascends a fishway twice as rapidly as expected, it will require a volume only half as long as expected, and the fishway capacity would then be double that expected.

Definite data on the rate of ascent of various species of fish through the varying types of fishways under varying conditions are needed, but unfortunately are lacking at the present time. Some research was carried out on the Columbia River on salmon and steelhead trout, but only a few results were published. These will be discussed in more detail in Chapter 3. As noted earlier, Jackson assumed a rate of migration of 5 min per pool for the conditions at Hell's Gate, which he felt was conservative. In illustrating a method used on the Pacific Coast in the *Canadian Fish Culturist*, September 1955, the author describes a rationalization based on the time that is required for a fish to swim a distance upstream through a height equivalent to the vertical distance surmounted through the fishway:

> The average distance the fish migrate upstream per day is obtained from biological tagging data. The vertical height the fish ascend per day is estimated by multiplying the distance travelled by the average river gradient. If the river rises three feet per mile and the fish migrate twenty miles per day, they climb through a vertical distance of 60 feet per day. Assuming they migrate through a twenty-hour period daily, they would climb three feet vertically per hour. If the fishway were designed to surmount a falls nine feet in height, it would therefore take the fish three hours to pass through it.

It will be seen that for the example quoted, the time required per baffle is 20 min, which is considerably higher than that assumed by Jackson. While this example is purely hypothetical, it is believed that in general this method will result in a slower rate of migration than will actually occur in a fishway. However, because the data on rate of migration from tagging are not likely to be available in many cases, it is expected that this method would have only limited use. It is felt that if a large background of research data and experience were available on migration rates in rivers and fishways, it might well be possible to correlate the two, so that for new fishways planned on a river system it would be possible to predict the rate of ascent through the fishway from the rate of migration in the river.

It should be remembered further, however, that for fishways larger than the foregoing minimum size, any variation in rate of ascent will affect the capacity and therefore the volume of the fish in direct proportion. While costs of fishways do not vary in direct proportion to the volume, there is definitely a relationship between the two, so that even a small variation in rate of ascent can have a significant effect on final costs.

Only one additional factor is needed in order to determine the fishway pool size, and that is the maximum number of fish the fishway will be required to pass in a given unit of time. Tagging data, or data from other sampling procedures may be available to give a picture of the distribution of the run and the maximum weekly, daily, or even hourly peak. In some cases this may be available from direct counts at a weir, while in others no data will be available, in which case it will be necessary to make an assumption if the rationalization is to be followed.

The decision whether to use the hourly, daily, or weekly peak for determining pool size will depend on the particular circumstances. It can be appreciated that the peak hourly rate might be considerably higher than the peak daily, which in turn could be much higher than the peak weekly rate. It is not possible at present even to suggest which should be used for varying species of fish in varying locations. However, it is possible to give some idea of what has been used. Jackson, in the case cited earlier, uses the peak hour rate, which was presumably deduced from tagging data. This, of course, was conservative, and based on the assumption that the peak hourly migration should not be delayed for a period of more than an hour. Proof was available that even short delays were harmful to Fraser River sockeye so there was ample reason for this conservative approach. In other cases, provision to pass the daily maximum has been considered acceptable, and there has been no reason to suspect that delays of less than a day, if they have occurred, have been particularly detrimental. If the maximum daily run is used as one of the criteria, still another assumption may be necessary to relate this figure to the rate of migration. The daily run must be converted to numbers of fish per minute in order to make use of the rate cited earlier of 2 to 5 min/ft of rise. Once again, in some cases actual counts or visual observations may have been made from which it is possible to state that migration takes place over a 16-h period or some other period each day, so that this computation can be readily made. If these data are not available, it may have to be assumed that the migration is distributed over the normal daylight hours.

When the three variables of space per fish, rate of ascent per baffle, and numbers of fish per minute are known or have been assumed, the resulting calculation to determine pool volume is quite simple. The following example may serve to clarify it even further, however:

1. Although other species may be present, the majority of the run is composed of fish with an average weight of 6 to 7 lb. Therefore the space requirement is assumed conservatively to be 4 ft^3 per fish.
2. From experience and judgment based on available data, the time of migration through a pool, with 1 ft difference of head between pools, is 3 min.
3. From tagging data and observations of concentrations on the spawning grounds, the maximum number of fish it is necessary to pass in one 16-h day is estimated to be 8000. This corresponds to slightly over eight fish per minute or about 25 fish per pool at 3 min per pool.

The volume of each pool at a minimum river stage would be $8 \times 10 \times 2 = 160$ ft^3. The volume required by the maximum day's run would be $4 \times 25 = 100$ ft^3, so the minimum size is adequate. If the minimum size had not been adequate, the pools could have been widened, deepened, or lengthened, until they met the volume requirement. In all probability, structural or space considerations would have decided which dimension to increase.

The selection of the slot width and other dimensions for a vertical slot baffle will follow from the pool size. It is the general practice to try to keep the average velocity in the pools to a maximum of approximately 1 ft/sec, as determined by dividing the discharge through the fishway by the cross-sectional area. For a pool 8 ft wide by 10 ft long, with a head drop of 1 ft per baffle, the slot width necessary has been found from experience to be approximately 1 ft. In most cases the velocity criterion of 1 ft/sec average will result in a fishway with little or no excessive turbulence. However, if there is any doubt, such as would be the case if a pool shape quite different from any previously used were desired, it would be wise to construct a hydraulic model of several pools of the fishway, at a scale as large as convenient, to determine the flow patterns likely to exist in the prototype. The importance of considering such model studies cannot be overemphasized, because it has been found that relatively small changes in baffle plan have resulted in completely unforeseen and undesirable hydraulic conditions in both the vertical slot and weir types of fishway.

For a Denil type of fishway the space requirement does not apply, since the fish on entering the fishway generally pass completely through without stopping. Zeimer (1962) estimated that the steep pass shown in Figure 2.5(c) had a capacity of 750 fish per hour. From the example quoted earlier, this would be adequate to pass the 8000 fish per day noted. Thompson and Gauley (1964) found rates of ascent of a steep pass up to 2520 fish per hour, while others found rates up to 1140 per hour. It should be noted that these latter rates were recorded in an experiment where headwater and tailwater conditions could be controlled. For natural obstructions it is recommended that the maximum figure as reported by Ziemer be used.

Denil fishways have seldom been used for surmounting obstructions more than about 12 ft (3 to 4 m) high in one run. For heights greater than this, the fish are provided with resting pools each 3 to 4 m of height ascended. The design of the resting pool would then follow the procedure noted for the pools of a vertical slot fishway already described.

Continuing with the vertical slot design, with the pool size decided and the fish entrance located by use of the profiles as described earlier, the number of baffles should be carefully checked and the location of the centerline also given careful consideration. To check the number of baffles, it might be desirable to construct a set of *headwater-tailwater curves.* To do this, the elevations of the water surface at the upstream and downstream ends of the fishway are determined from the water-surface profiles at various rates of discharge. These are plotted, and smooth curves are drawn through the points as shown in Figure 2.10. The operating limits of the fishway can also be shown. It is then a simple matter to determine the maximum distance between the two curves and ensure that enough baffles are placed in the fishway to overcome this head. This curve can be particularly advantageous where some headwater or tailwater elevations are missing. Extrapolation of the curve within reasonable limits will often reveal the possibility of unexpected conditions of head, which it may be desirable to consider in the design of the fishway. As would be expected, the curves of headwater and tailwater are seldom parallel or straight lines. Good judgment and

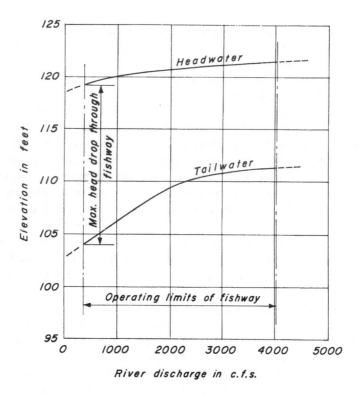

Figure 2.10 **Headwater-tailwater curves for the fishway constructed at the site for which the hydrographs and the water-surface profiles are shown in Figures 2.7 and 2.9.**

some boldness are therefore desirable, both in their construction and their application, particularly where data are scarce.

In finally locating the centerline of the fishway, the structure with the required number of baffles can be tried in any desired number of locations, with the downstream end fixed at the point where fish must enter. The location involving the least amount of rock excavation will often be easily determined, but it may not always be the best in other ways. For instance, it may be desirable to move the centerline or even the entrance a few feet in order to make the proposed fishway walls line up with existing fissures or planes of weakness in the rock. This will help to prevent overbreak and make the rock excavation easier. A geologist can be very helpful in this respect as he or she is often able to predict subsurface conditions, such as cracks, from surface observations, and the engineer can be guided accordingly. In addition, it may be found more economical in some cases to place the fishway in a tunnel excavated in the bank rather than in a deep open cut. If this appears to be the case, the geologist can assess the stability of the tunnel roof and indicate the degree of care needed in blasting. The same considerations apply to the siting of a Denil fishway.

The best practice is first to try to arrive at the most economical layout of the fishway on paper, meeting all hydraulic criteria, and then to mark this out on the ground at the site. At this stage the geologist can inspect the layout and advise whether any changes are desirable, and if the engineer is also present, these changes can be settled on the ground.

The depth of water in the fishway will be governed by the range of discharges occurring during migration. Here again, the headwater-tailwater curves may be used to select the desired range of elevations for the structure. If the tailwater fluctuates over a larger range than the headwater, it may be necessary to make the structure deeper at the downstream end than at the upstream end, while in other instances the reverse might be necessary. To meet the volume criteria discussed earlier, the fishway should be designed to have a minimum assured depth of 2 ft (60 cm) of water during migration periods. If it appears likely that this depth will occur frequently during migration, the minimum should be increased, as 2 ft of depth is hardly adequate from the point of view of swimming depth, particularly when the fishway is crowded. Normally the minimum depth would occur only rarely, and the ideal depth of about 6 ft can be designed for times when peak migration is likely to occur.

Again, for a Denil fishway such as the steep pass, different criteria would apply. Since it operates up to a maximum depth of only approximately 6 ft (2 m), it could be sited only where the range of flows would give depths in it of 2 to 6 ft (0.6 to 2.0 m).

In some cases it is possible to locate the centerline so that the fishway is entirely in rock cut for its full depth. If the rock is sound and will break cleanly with proper care in blasting, it will be possible to dispense with formal concrete fishway walls and build the baffles between the rock walls. In most cases, however, it will be found that low walls are required to prevent the river from flowing into the fishway at some point or points along its length. These places can be easily determined by reference to the water-surface profiles and the topographic details. By plotting the plan of the profiles on the topographic plan showing the contours and the fishway centerline, it can readily be seen where this condition will occur.

2.10 DESIGN OF FISHWAYS FOR NATURAL OBSTRUCTIONS — STRUCTURAL DESIGN AND CONSTRUCTION

Having selected the functional dimensions and the location of the fishway, and having determined the wall heights needed, the structural design can proceed. Selection of construction materials will depend largely on availability, costs, life expectancy of structure, and a number of other factors associated with the particular site and the program under way. While in most cases reinforced concrete is more economical than steel, in some cases no sources of aggregate may be found near the site, and the cost of transporting it over long distances may outweigh any advantages to be gained by the use of concrete. For Denil fishways the possibility of precast concrete, steel, or aluminum plate might be considered. No definite set of rules can be laid down as to choice of materials. It is left to the ingenuity and experience of the designer to select the materials best suited to local conditions.

The essential structural elements of a typical vertical slot baffle for small fishways are shown in Figure 2.11. In some designs the center column A, pilaster B, and baffle plate C have been constructed of structural steel fastened to the concrete walls and floor of the fishway. Other designs have called for the first two of these members to be made of reinforced concrete, with timber for the baffle plates, or with

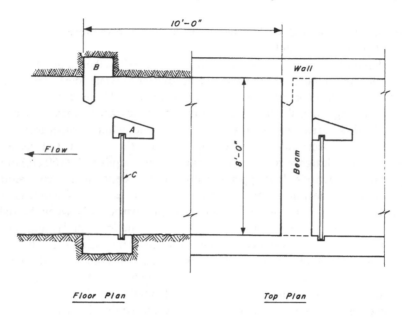

Figure 2.11   Plan showing the structural elements of a small vertical slot fishway.

all members made of concrete. In still other designs, members A, B, and C have been constructed of pressure-treated timber.

Any structural design starts with a determination of the loads that can be expected to create the maximum stresses in the structure. The worst condition to be expected in a fishway occurs when the slot in one baffle becomes completely blocked by debris.

This condition could impose the full head difference between head- and tailwater on the blocked baffle, acting in a downstream direction. There is a reasonable doubt, however, whether this condition is likely to occur frequently enough to warrant design to such stringent standards. In many cases, particularly for smaller fishways, this maximum loading can be economically designed for, because other criteria, such as steel reinforcing needed for temperature stresses, are adequate to meet it.

The design of beams and columns, etc., to meet the loading specified is a straightforward procedure, and any good structural design text will provide necessary advice in this respect. As noted earlier, it is often possible to economize by dispensing with formal walls inside the rock cut, provided that the rock conditions are satisfactory after excavation. The bare rock walls do not detract hydraulically from the functioning of the fishway, even if overbreak well outside the line of the fishway walls has occurred in blasting. If the rock is not sound, and is subject to erosion later, it may be desirable to line the rock walls with concrete. If the intention is to use the rock walls without lining, particular care should be taken in the method of drilling, blasting, and excavating. Drill holes should not be too long, and the center portion of the rock cut should be drilled and blasted before the portions next to the sides. This provides a relief area in the center of the fishway to receive the material blasted away from the walls, and eliminates a large part of the shattering effect of the explosion on the finished rock walls. The same care must also be used where the fishway is to be placed in a tunnel.

If concrete walls and floor are to be used in the fishway, the baffle members can be designed to span between them in a number of different ways. If the bare rock is used as a finished wall, the end connections of the baffle members have to be given particular consideration. There are a number of ways they can be connected to the rock, but experience has shown that the most satisfactory way is to insert a reinforced concrete beam in a trench excavated across the floor with its top surface flush with the floor, and to place the columns similarly in the walls at each baffle. The other baffle members can then be connected to these frames monolithically if concrete is used or by bolting or direct anchoring if steel members are used. If the beams and columns set flush in the floor and walls, respectively, are not securely held in place because the rock is not sound, it may be desirable to anchor them with steel dowels. These can be drilled and grouted into the nearest solid rock to provide stability. The need for these will depend on the particular conditions occurring at each baffle location. Such measures may be needed at one baffle and not at another, and it might be impossible to predetermine where they will be needed. In such cases, inspection of the completed excavation is necessary before deciding on the need for dowels at each baffle.

The provision of a grating cover for the fishway, of either concrete slab or steel mesh, is optional, but it has been found to be useful to protect the baffles in areas where erosion of the bank above the fishway occurs, or where the river carries a heavy bedload or much floating debris during freshets (see Figure 2.12). Steel trash racks are designed for the full head resulting from complete blockage. The trash rack bars must have wide enough spacing to permit free passage of the fish between them.

2.11 MAINTENANCE AND EVALUATION OF FISHWAYS AT NATURAL OBSTRUCTIONS

While maintenance and evaluation are not related in purpose, they are dealt with together here because they are both necessary and important steps that should be taken following completion of a fishway.

Adequate maintenance is an accepted requirement in good engineering practice, and is vital to successful operation of a fishway. Fishways should be inspected at least once a year and repaired as often as necessary to ensure continuous operation throughout the life of the structure. Because fishways are often subject to extreme conditions of temperature, to violent fluctuations of river flow, and to debris loads of considerable magnitude, annual maintenance costs are high, amounting to as much as 5% of the cost of the structure. The type of maintenance normally required consists of straightening or replacing steel members in the trash racks, baffles, or gratings; patching or replacing concrete lost through abrasion by bedload, or by spalling; repainting structural steel members; and removing bedload and debris. Painting is required frequently if a structure is inundated each year to considerable depths, because the moving bedload rapidly abrades even the toughest paint from exposed surfaces.

Evaluation of the use of a completed fishway is more difficult than evaluation of the use of most engineering structures. Use of a new bridge or a new passenger elevator in a building can easily be assessed by simple traffic counts. The use of a fishway can also be checked by count, but aside from the difficulty of obtaining the count, it does not always tell the full story. Turbulence and turbidity often prevent

**Figure 2.12 Pictured on the left are the steel trash racks on the fish exit of the Left
Bank High Level Fishway at Hell's Gate. The widely spaced bars protect
the fishway while the closely spaced bars protect the intake to the auxiliary water supply. The insert, lower right, shows the steel gratings on the
fishway constructed at the site shown in Figure 2.10.**

satisfactory visual counts in fishways at natural obstructions. In addition, in many
cases a proportion of the fish may still be attempting to surmount the obstruction
directly without the benefit of the fishway, some of them successfully, while others
drop back to perish or be delayed before passing through the fishway. In most cases

a tagging experiment is the only way to determine how successfully the fishway is passing the run. If possible, this experiment should duplicate one undertaken before construction so that a direct comparison can be made. For example, for a small fishway on the Pacific Coast of Canada, it was found that the average delay for tagged fish at the obstruction was reduced from 14.6 to 1.7 days by construction of the fishway. This kind of proof is encouraging in any program of elimination of natural obstructions to fish migration.

In some cases the indirect proof of the success of a fishway is so overwhelming that no further tests are necessary. At Hell's Gate, where the vertical slot fishways were used for the first time, the increase in the populations spawning above the obstruction has been spectacular, and no further tagging experiments have been necessary. It may be obvious in many cases that fish are ascending in large numbers to areas that were only sparsely populated before, so that proof by tagging is not necessary. Whatever the method, some evaluation of use of a fishway should be made after construction to ensure that the structure is serving its purpose. This evaluation is not an economic one, because economic benefits in increased production will not be evident in most cases for many years. Rather, the evaluation of use determines whether the structure is functioning properly.

2.12   LITERATURE CITED

Bell, M.C., 1984. Fisheries Handbook of Engineering Requirements and Biological Criteria, U.S. Army Corps of Engineers, North Pac. Div., Portland, OR. 290 pp.

Committee on Fish Passes, 1942. Report of the Committee on Fish Passes, British Institution of Civil Engineers, William Clowes and Sons, London. 59 pp.

Denil, G., 1909. Les echelles a poissons et leur application aux barrages de Meuse et d'Ourthe, *Annales des Travaux Publics de Belgique.*

Jackson, R.I., 1950. Variations in flow patterns at Hell's Gate and their relationships to the migration of sockeye salmon, *Int. Pac. Salmon Fish. Comm. Bull.,* 3(Pt. 2), pp. 81–129.

Katopodis, C. and N. Rajaratnam, 1983. A Review and Lab Study of the Hydraulics of Denil Fishways, Can. Tech. Rep. Fish. Aquatic Sci. No. 1145. 181 pp.

McLeod A.M. and P. Nemenyi, 1939–1940. An Investigation of Fishways, Univ. Iowa, Stud. Eng. Bull. No. 24. 63 pp.

Pretious, E.S. and F.J. Andrew, 1948. Summary of Experimental Work on Fishway Models, Int. Pac. Salmon Fish. Comm., unpublished data. 6 pp.

Thompson, C.S. and J.R. Gauley, 1964. U.S. Fish & Wildlife Serv. Fish Passage Res. Prog. Rep. No. 111. 8 pp.

Zeimer, G.L., 1962. Steeppass Fishway Development, Alaska Dept. Fish & Game Inf. Leafl. No. 12. 27 pp.

3   FISHWAYS AT DAMS

3.1 DAMS AND THEIR EFFECTS ON MIGRATORY FISH

It must be made clear at the beginning that the mere provision of fishways or fish-passage facilities at a dam does not ensure the continued existence at their original level of an abundance of the migratory fish for which the facilities were designed. The construction of a dam in a stream can have many diverse effects on the physical characteristics of the river. Water temperatures can be changed both above and below the dam. The normal pattern of seasonal flow in the river can be altered, with floods occurring later than normal or not at all. Silt may be deposited in the reservoir above the dam, and natural riverine spawning grounds can be flooded. Water with a very low oxygen content may be released below the dam. All these physical changes can greatly affect, either directly or indirectly, any fish passing upstream or down through the altered portion of the river. Direct effects might consist of lethal or sublethal temperatures where normal temperatures existed previously. Indirect effects would be any change in food supply and disruption of the ecological balance. Any of these effects could pose serious problems, and their solution could be as vital to the continued existence of the fish as the construction of fish facilities.

The fishway, then, is the answer to only one of the many problems created by the dam. In many cases, however, the other problems may be minor and thus be ignored. This would apply in the case of a low dam on a stream populated with comparatively few fish. In general, as the number of migrating fish increases, its effect on the physical environment increases and adds further complexity to all the problems created by the dam.

The discussion of fishways and other fish-passage facilities in the following chapters should be regarded in the light of the foregoing comments. In many of the examples quoted, the provision of fishways has been the only measure taken to answer all the fisheries problems associated with the dam. While the fishway may have solved the more obvious problem, it has not been proved in most cases whether this measure alone has been adequate. It takes many years to determine the trend in a fish population, and often, when a downward trend eventually has been shown, its cause has been obscured by a number of different effects that have not been segregated and evaluated.

It is hoped that research will eventually clarify the complex picture of fish migration past dams and enable more complete measures to be taken to preserve the fish when most dams are built. In the meantime, progress must be made largely by trial and error, and it is hoped that the following, which is a record of practice and experience in various locations under varying conditions, will be of some value in this regard.

3.2 TYPES OF DAMS

It is necessary to know something about the types of dams that can be constructed and their purpose, in order to approach the design of fish-passage facilities with confidence. Dams can be subdivided very generally into those that impound water for release as needed, and those that merely divert water for use elsewhere. Some dams may combine these two purposes. The dams that impound water may be further classified into those that impound water for release into the original stream for use at a point some distance downstream (known as storage or flood control dams) and those that impound water for use at or near the dam for hydroelectric-power production or for some domestic or industrial use. In the latter case the height of the reservoir surface is important, as is the quantity of water stored, because this height, or *head* as it is called, is directly proportional to the power potential and to the usefulness as a domestic or industrial water supply. Dams have also been built for a number of other purposes, such as providing slack water for navigation, providing waters for recreational purposes or for rearing fish, or maintaining the water table for irrigation. However, the uses outlined previously are the most common.

Dams may also be classified according to their design, particularly as it affects the materials used in construction. The earliest dams built by humans were probably weirs constructed of brushwood and earth, similar to beaver dams. Later, dams were constructed of timbers securely fastened together, or of masonry. Actually, masonry dams (without mortar) were built for irrigation and water supply in ancient times, but several thousand years elapsed before the modern dam-building era was born. The building of dams gained real impetus when humans began to use water to generate power, first by direct drive from a water wheel to the stones to grind grain, and later by the generation of electricity. Dam design and construction in modern times has kept pace with the rapid expansion in use of electricity, even though many of the dams have been built for other purposes as well. While dams may occasionally still be built of timber or masonry, most dams today are constructed of concrete or earth (or rock) fill. Concrete dams may be of arch design, in which case they are comparatively thin, or of gravity design, in which case they are rather thick, depending on weight alone for their stability. They may also be of a design known as the "Ambursen," or buttress, dam, which consists of a series of frames on footings with thin concrete plates spanning between them. Rock fill dams have side slopes with a relatively steep angle of repose. Earth fill dams, on the other hand, have to be built with much flatter side slopes, resulting in a dam with a very thick base.

Any of the foregoing types of dams may be used for either impounding or diverting water, or for any of the uses listed previously, although some are better suited than others for particular uses. It would be well to review in more detail some of the types listed and perhaps indicate how the design of the dam might affect the provision of fishways. (See Figure 3.1.)

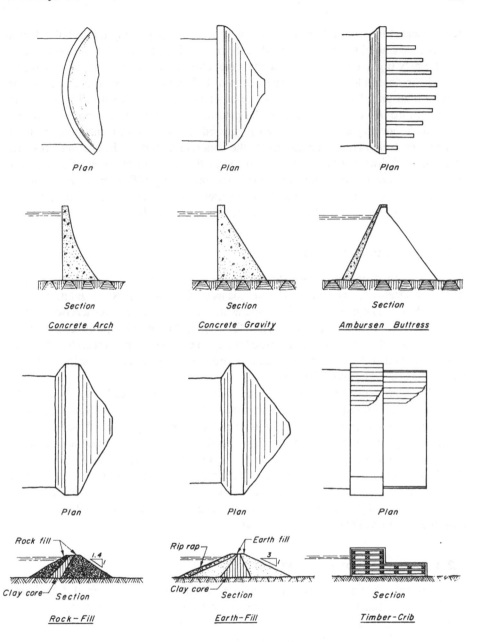

Plan   *Plan*   *Plan*

Section   *Section*   *Section*

Concrete Arch   Concrete Gravity   Ambursen Buttress

Plan   *Plan*   *Plan*

Rock fill   Earth fill   Rip rap

Section   *Section*   *Section*

Clay core   Clay core

Rock-Fill   Earth-Fill   Timber-Crib

Figure 3.1 Plans and cross sections of several types of dams.

3.2.1 Concrete Arch Dam

This type of dam is well adapted to narrow rock canyons, where the arch thrust can be readily transferred to the solid rock walls. Figure 3.1 shows a typical plan and cross section. The thin section of the dam will be noted as well as the steep downstream face, and it will be appreciated that incorporation of a fishway in such a dam would be difficult. Usually a fishway at this type of dam would be most economical if it were built into the canyon walls, merely passing through the dam at the necessary height to release the fish at the reservoir surface. Often this type of dam is extremely high, and situated on streams that carry small runs of fish, in which case it might be desirable to forego construction of fishways in favor of trucking the fish upstream around the dam or using some other type of fish elevator. Fishways that spiral upward in a circular or rectangular pattern in the manner of ramps in a modern automobile parking garage have been suggested as a solution for installation on the steep downstream face of dams such as the concrete arch dam. In practice, this idea is not new; fishways have been constructed in this form in many existing installations. However, the economics of a spiral fishway, rising 100 m or more, are highly questionable. In some cases, the concrete arch may be combined with a gravity section, but in most cases the same comments would apply.

3.2.2 Concrete Gravity Dam

This is one of the most common types of dam, used extensively for hydroelectric projects on the main stems of larger rivers. The great dams on the Columbia River and many of the dams on Russia's large rivers are of this type, as are those on the Nile River at Aswan, Egypt, and the Volta River in Ghana. The design of fishways at dams of this type has become fairly standardized on the Columbia River, where fishways have been provided at each of the many dams below Grand Coulee Dam. These dams are from 50 to 100 ft high, and the fishways have been constructed integrally with them.

When dams are constructed in series, so that each is located at the head of the reservoir formed by the next dam downstream, the powerhouse is usually located at the dam and is constructed integrally with it. Because a large part of the river flow passes through the powerhouse at times, and fish migrating upstream are naturally attracted to this flow, an arrangement for collecting fish from this flow has been developed, which has become known as the *powerhouse collection system*. Because of this feature and others of equal interest, all of which are designed to pass large numbers of valuable fish upstream, this type of dam and fishway installation will be described in more detail later in this text.

3.2.3 Ambursen, or Buttress, Dam

This dam is particularly well suited to locations where special foundation problems exist but is not necessarily limited to such locations. It is a concrete dam that can be constructed on material other than a rock foundation. Because dam designers prefer a rock foundation where possible, this type is not common. The upstream face of this type of dam is a sloping slab, and the dam depends in part for its stability on the weight of water acting on this upstream face. Since it is constructed of concrete, there is no reason why it cannot be provided with fishways similar to those used for the gravity dam.

3.2.4 Earth and Rock Fill Dams

Earth and rock fill dams are constructed by building a dike of carefully graded materials across the river. Usually an impervious central core of concrete or clay is provided, which forms an effective water seal down to bedrock where possible. The fill materials are built up in layers on either side of this central core to required height, as shown in Figure 3.1. It will be seen that the result is a dike-like structure of considerable thickness at the base. Since the flatter slope is used for earth fill dams as compared to rock fill, the earth fill dams are extremely thick at the base. This type of dam is characterized by settling as it ages, so that it is not practicable to build a concrete spillway over its top, and dam designers are reluctant to place any other type of outlet works either through or over it. The spillway is therefore usually placed in rock on one of the river banks adjacent to the dam, or at some more distant point along the shoreline of the reservoir, where rock foundation exists and overflow back to the original river can be accomplished.

The problem of fish passage will occur where the spillway and powerhouse are located, of course, rather than at the dam. There is no obvious reason why fishways cannot be constructed along with or adjacent to the spillway. If a powerhouse is involved, however, it could be at some distance from the reservoir and a fishway might not be economical under these circumstances. If the run is not too large, it might be well in such a case to consider another method of passing fish such as hauling by truck. Earth and rock fill dams are being used more extensively, and are being designed for increasingly greater heights, as time progresses. They have advantages in economy in those locations where materials are readily available near the site, and undoubtedly their use will continue to increase in the future.

3.2.5 Timber Crib Dam

This type of dam was very common in North America in the days when timber was plentiful. It is now almost nonexistent even in areas where timber is still available in quantity. Some are still being built, however, particularly where their use is only temporary. Timber cribs are often used for cofferdams for river work, because they are easy to place by simply floating them to the site and sinking them, and are very easy to dismantle after they have served their purpose. Timber crib dams depend mainly on the weight of the rock placed in them for stability and are usually broad crested. It is comparatively easy to fit a timber fishway through such a structure, and in areas where timber is available, these fishways can be quite economical. Frequent maintenance is usually required, and maintenance costs are probably higher for these than for other types.

3.3 HOW DAM OPERATION CAN AFFECT FISHWAY DESIGN

A number of complex and unexpected problems can be presented to the fishway designer as a result of the operation of a dam. These must all be taken into consideration in the design of the fishway, insofar as it is possible to foresee them.

First of all, for a hydroelectric dam, it should be known whether the plant is to be operated only periodically at times of peak load, or continuously to produce firm power, or as a combination of these depending on the load requirements at any given time. The designers and operators of the plant can provide this information, but it is often possible to judge for oneself by comparing the water available to the plant requirements, and then simply determining if it is likely that there is going to be enough water to keep the plant running continuously at anywhere near its full capacity or peak efficiency. If the plant is to operate for peaking purposes, delay to the migrating fish is possible because the only attraction to the tailrace area would be the small flow from the fishway at times when the plant was shut down. It might be necessary in such cases to provide a small flow through one of the turbines to assist in attracting the fish, or to take other measures to ensure a continuous movement of fish. Conversely, if the plant is operated as a base load plant, the reverse can be true, with a continuous flow from the powerhouse and large seasonal flows over the spillway. In this case the fish could be attracted to the continuous powerhouse flow, or to the large spill area at the base of the spillway if they are migrating at a time of year when a surplus of water exists. Combinations of these two types of operation will produce complex patterns of flow release between the powerhouse and spillway, and these should be studied and predicted as far in advance as possible in order to determine the most advantageous location for the fishway entrance.

In Scotland, where hydroelectric plants are frequently operated for peak load periods only, the condition where plants are shut down completely for large parts of the day is common. As a consequence, to provide transportation water for the upstream migrant salmon and also make the river suitable for angling, the owners are required to release what is known as *compensation water* in amounts determined by the fisheries authorities or local river boards. This compensation water is usually a substantial quantity and can be several times the quantity required to operate the fish facilities. In many cases the release of this water on a continuous basis is arranged in such a way as to provide attraction to the fishway entrance. Fishway entrances will be dealt with in more detail in a later section, but this is one instance of the often complex nature of design and operation problems encountered.

The operation of a hydroelectric power plant can also affect flow in the fishway. If the reservoir is small, sizable daily fluctuations could be expected in the elevation of its water surface. The effect of these rather rapid fluctuations on water flow in the fishway has to be considered in design. Automatic adjustable weirs might be necessary to ensure a uniform flow of water into the fishway throughout the day. The same thing could apply to the fish entrance and lower end of the fishway. Substantial fluctuations in tailwater level might occur, which would back up into the entrance and lower end of the fishway each day, thereby reducing the entrance velocities and making them less attractive to the fish. Besides these daily fluctuations, seasonal fluctuations in headwater and tailwater must be anticipated in the fishway design. These may have some resemblance to normal seasonal variations, but they could also be changed considerably by the operation of the power plant and by auxiliary storage works.

For a storage dam integrated with a hydroelectric development downstream, daily fluctuations of spill to meet load requirements are possible, but it is more likely that steady releases will occur since they require less frequent adjustment. However, these releases are likely to have a very different pattern than normal river flows, and any fishway associated with the storage dam must be capable of functioning properly under all combinations of headwater and tailwater levels likely to occur when fishway operation is required.

3.4 THE FISH ENTRANCE — GENERAL

Probably the most important single part of any fishway, particularly those at dams, is the fish entrance. If the entrance is not readily found by the migrating fish, they will be delayed for varying lengths of time, and in the extreme case, they may never enter the fishway and migrate upstream. The reason that the fish entrance is even more important at dams than at natural obstructions follows logically from the argument raised earlier in the introduction to Chapter 2. Any dam that forms a barrier to the upstream migration of fish imposes an entirely new stress on the fish that normally migrated upstream past the site prior to construction of the dam. This stress is not limited to the effects of enforced delay, but includes the effects of all the other physical changes in the environment of the fish (temperature, velocity, water quality, etc.) resulting from the construction of the dam. While the effects of these latter changes can be very important, we are concerned here only with the stress resulting from overcoming the physical barrier itself. If this new stress is caused by delay only, it is essential that it be reduced to a minimum, or for all practical purposes eliminated, if the fishway is to be effective. Even though the fish it is desired to pass can tolerate some delay, the sooner they enter the fishway, the more effective and efficient it will prove to be in the long run.

If it were possible to artificially shape the banks of a river below a dam so that in plan they converged to the fishway entrance, it would be a simple matter to lead the fish into a fishway. Alternatively, if the fishway entrance could be widened to the full width of the river below a dam, the fish would have no difficulty in finding it, and delay could be minimized or avoided. Impractical as it sounds, this has actually been accomplished in some cases. In one case in particular on the Okanagan River, (see Clay, 1960) the entire spillway at each of 13 low dams was made into a fishway so that migrating fish had no difficulty in finding a route upstream (see Figure 3.2). Another case could occur at hydroelectric projects where river flow is sufficiently controlled so that no spill takes place when fish are migrating. At such projects, where an efficient powerhouse collection system is installed, the fish are inevitably attracted to the area of the fish entrance because all the river flow is coming from this area. Such a collection system will be detailed later in this section.

A multiplicity of layouts is possible in the arrangement of the spillway, powerhouse, water intake, and nonoverflow section at dams. The topography and foundation conditions at the site usually determine the final layout chosen. It would be impractical to consider here all the possible combinations of conditions that might be met. We will start off therefore by considering only a few simple layouts that demonstrate some of the principles involved in designing the fish entrance, and later on we will refer to a few more examples that demonstrate problems of interest.

Figure 3.3 shows three schematic plans of simple dam arrangements. Plan A is typical of low weirs or dams used for irrigation diversions, power diversions, navigation, etc. It is assumed in this plan that any diversion canal or pipe is some distance upstream from the dam. Plan B is a similar type of layout except that there are control gates on the spillway rather than a free crest as in A, and the channel has been narrowed below the dam to little more than the width of the spillway. Plan C is similar to B, except that a powerhouse has been incorporated in the dam on the left bank.

Figure 3.2 **One of 13 low dams on the Okanagan River in British Columbia, where the entire spillway was stepped down to form fishways extending over the full width of the river.**

Figure 3.4 shows schematically some typical cross sections of dams that illustrate the position of the upstream limit of migration under the most commonly occurring conditions. Cross section A shows the condition at a simple low weir with a free fall over the crest. This condition is not illustrated in the plan in Figure 3.3, but it can be visualized by removing entirely the shading denoting the upstream limit of migration in Plan A, so that the upstream limit becomes the face of the dam. This condition is fairly common at low weirs where the depth of flow over the weir is not great. It will be described in more detail in Chapter 5.1 dealing with barrier dams, where it is often attempted to achieve this condition. Cross section B of Figure 3.4 shows a common type of Ogee spillway with dimensions and flow such that the phenomenon know as *hydraulic jump* occurs. This is characterized by a sudden rise, or "jump," of water to a level higher than that at the toe of the spillway. The water at this point recovers some of its static energy in exchange for the kinetic energy it developed in falling over the dam. This condition is indicated in all the plans in Figure 3.3, where the line showing the upstream limit of migration corresponds to the line in the cross section at or near the top of the hydraulic jump. Cross section C shows a typical large installation housing a comparatively low-head turbine of the propeller type. It will be seen that fish approaching the powerhouse are able to

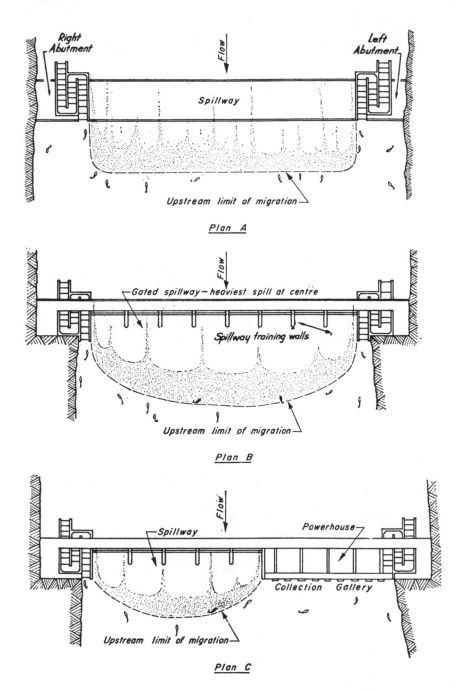

Figure 3.3   Three schematic plans of simple dam and fishway arrangements: (A) A free crest spillway; (B) a gated spillway; (C) a combined gated spillway and powerhouse.

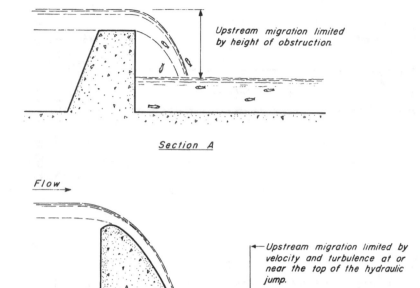

Upstream migration limited by height of obstruction.

Section A

Flow

Upstream migration limited by velocity and turbulence at or near the top of the hydraulic jump.

Section B

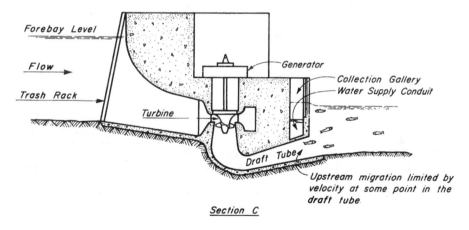

Forebay Level

Flow

Trash Rack

Generator

Collection Gallery

Water Supply Conduit

Turbine

Draft Tube

Upstream migration limited by velocity at some point in the draft tube.

Section C

Figure 3.4 Typical cross sections illustrating the position of the upstream limit of migration below dams.

penetrate the draft tube only a short distance and will have to sound the required depth to do so, but near the surface they are blocked by the downstream face of the powerhouse structure, and it is here that they can most conveniently be attracted into the collection gallery shown.

From the two sets of diagrams in Figures 3.3 and 3.4, it will be seen that fish approaching a dam as they migrate upstream will be met by a barrier that can be formed by high velocities and turbulence occurring in the spillway bucket or in the draft tube, or by the downstream face of the dam or powerhouse. Over the years, experienced designers of fish facilities have frequently commented that the entrance to a fishway should be as close as possible to the point or line to which the migrating fish penetrate farthest upstream at an obstruction. Strangely enough, this is still one of the most frequently overlooked criteria in fishway design. According to theories on the physiological process that guides salmon and other anadromous fish back to their home streams, it is the combination of an odor response and the behavior pattern of swimming upstream against the velocity that enables the fish to return to their place of birth. They will not return downstream unless the odor of their home stream is missing. Therefore, at a dam where the water above and below has the same constituents, the fish will always continue to work their way upstream, so that there is every reason to believe that this maxim of fishway design is still as correct as it ever was, and should be adhered to at all costs.

Merely placing the entrance at the farthest point upstream is not the entire answer to the problem, however. It will be seen from Figure 3.3 that a variation has been incorporated in Plan B, which tends to make the upstream limit a barrier that guides rather than merely stops the fish as in Plan A. A further change in Plan B will be noted in that the eddy downstream of the abutment on each bank has been eliminated. The reasons for these changes will receive further elaboration in the ensuing section.

3.5 FISH ENTRANCES — SPILLWAYS

The layout shown in Plan A, Figure 3.3, even though it has two fish entrances placed at the farthest point that fish can reach when proceeding upstream, is not desirable for two reasons.

First, if the dam were constructed across a wide river, and it was possible and could be expected that fish would approach it at points all across the river, then fish approaching it near the middle of the river would have no encouragement to seek the banks for a possible route upstream. Any small movement from side to side would place them no farther upstream, so that not until they became desperate and had to widen their search for a route upstream could they hope to find and enter the fishway. Similarly, if on a narrower river it is decided to construct only one fishway, for some reason such as economics, fish approaching the dam along the bank opposite to the fishway have no guide to encourage them to move across to the other bank, where the fishway is located.

An attempt to overcome these disadvantages has been made at some dams by the adjustment of the flow through the spillway gates. To make use of this method, it is necessary, first, that there be gates on the spillway, second, that there be a sufficient number of separate, independently operated gates to be able to produce the required flow pattern, and third, that full spillway capacity is not required at the time fish are migrating. Assuming that these requirements can be met, it is possible, by opening wide the spillway gates at the center of the river and gradually decreasing the opening at successive gates toward the shore, to form a pattern something like that shown in Plan B, Figure 3.3. This pattern, as may be seen, tends to guide the fish toward the fishway entrances, on the basis that their behavior is such that they swim against the velocity barrier and continue to work upstream simultaneously. Similarly, with a fishway on one bank only, it is possible to manipulate the gates so that the heaviest spill is at the bank opposite the fishway, and the smallest spill is adjacent to the fishway, with the result that the velocity barrier forms a diagonal lead across the river to the fish entrance.

While this arrangement sounds almost ideal, its practical application involves many difficulties. Some of these difficulties stem from the fact that the position of the hydraulic jump varies with the spillway discharge. Most spillways are designed to get rid of excess flows as they occur in the river, with the result that quantities spilled are seldom constant for any length of time. Besides the prodigious task of continual adjustment of gates to secure the required pattern under frequently changing flows, a problem occurs in the spillway bays adjacent to the fishway. Here, as the flows decrease, the hydraulic jump can either move back into the bay between the training walls or disappear altogether, permitting the fish to enter the spillway bay far enough to be trapped between the training walls. On the Columbia River dams where this problem has occurred, a device known as a *picketed lead* has been used in an attempt to alleviate the problem. This is a steel trash rack with bars spaced closely enough to stop fish ascending farther. It is placed across the spillway bay adjacent to the fishway entrances shown in Figure 3.5. This arrangement in effect adds the comparatively large volume of flow from this adjacent spillway bay to the attraction flow issuing from the fishway itself. This solution has not been entirely satisfactory, however, because of excessive trash accumulation on the picketed leads, which results in a heavy maintenance problem. Where picketed leads have been used, a velocity of 1 ft/sec through them has been adopted as the best compromise between practicality and efficiency.

Other difficulties stem from the fact that spillway capacity is limited by the necessity of setting up the desired flow pattern. In many cases it is not practicable to gradually reduce flows at each successive spillway gate from the center of the river toward each bank. The best and possibly only compromise is to operate all the spillway gates the same with the exception of the gate adjacent to the fishway entrance, which is operated in conjunction with a set of picketed leads as described. This does not result in the ideal flow pattern shown in Figure 3.3, Plan B, but it does result in a better guiding pattern than that shown in Plan A.

The other reason that the layout shown in Plan A is not good is the undesirable circulation set up in the area downstream of the dam abutments, which project some distance into the river. This can be avoided by excluding the abutments from the channel downstream, as in Plan B, but this is not always an acceptable or practical solution from the point of view of the designer or owner of the dam. It is therefore necessary to understand why this condition is undesirable in order to provide the best arrangement possible if, as is normally the case, some compromise is necessary.

Figure 3.6 is an enlarged plan of the right bank abutment. The circulation of the water in the eddy is shown by the small arrows. It will be noted that flow along the shore is actually in an upstream direction, so that fish proceeding in their normal direction against the current, would actually be headed downstream if they happened to enter this shoreline area. It is quite probable that they will enter this area,

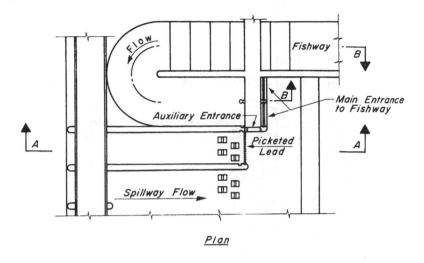

Plan

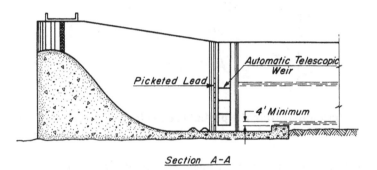

Section A-A

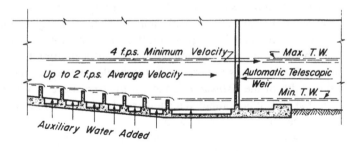

Section B-B

Figure 3.5 Typical spillway fishway entrance with picketed lead as used on the Columbia River dams by the U.S. Army Corps of Engineers.

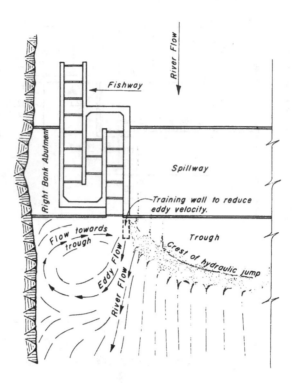

Figure 3.6   Plan illustrating the probable direction of currents below the right-bank abutment of a typical dam with fishway adjacent to spillway.

particularly if high velocities in the center of the river force them to swim upstream along the margins. It could be argued that if this is a circular eddy as indicated by the diagram, and the fish proceed against the current, they will eventually pass in close proximity to the fishway entrance and have the opportunity to enter it. It must be remembered, however, that they will probably approach the fishway entrance tail first, and the fact that the eddy may be large means that the comparatively small flow coming from the fishway is not able to compete successfully in attracting the fish. The size and volume of the eddy is important. If it is small, say, equal to or smaller in breadth than the width of the fishway entrance, it may not have too harmful an effect. If it is larger than this, it is a potential hazard and should be thoroughly studied to determine if it can be eliminated or reduced.

It is not always possible to predict by looking at the dam plans whether an undesirable eddy will exist after the dam is built. If it appears possible that an eddy will occur, it should be eliminated as far as possible by narrowing the channel as in Plan B, Figure 3.3. If this is not possible, and economics warrant it, a hydraulic model should be constructed and various solutions tested. Such a model need not be expensive, because a model of only the immediate area including a small part of the spillway is often adequate.

The cause of this eddy is usually the difference in static head between normal tailwater elevation near the shore and the lower water-surface elevation at the upstream end of the hydraulic jump at the base of the spillway. This difference in elevation causes a velocity from the shore toward the base of the spillway, where the water is at its lowest level. The circulation is further encouraged by the drag of the high downstream velocity of the main river flow on the circumference of the eddy. This sets up the clockwise flow in the eddy below a right bank abutment as shown in Figure 3.6, which would, of course, be reversed if it occurred on the left bank. It is often possible to damp this eddy by inserting a short training wall between the fishway entrance and the spillway bay as shown in the figure. There is a limit to its possible length, however, because if it is extended too far downstream, fish could be trapped in the spillway bay.

The same eddy usually exists at dams where there is no hydraulic jump (Section A, Figure 3.4). In this case it may be due to the static head difference in the same areas, resulting from a lowering of the level of the water between the jet and the dam by the action of the jet itself. It is also possible to damp circulation of this eddy by use of a stub wall. In some cases, however, the space between the falling water and the face of the dam can be utilized as a fish passage. If this is the case, it would not be desirable to block the circulation off with a stub wall, but it would still be desirable to prevent any large eddy from circulating between the fish entrance and the bank. The use of the space between the falling water and the dam for fish passage is also noted in Chapter 5.

If the eddy can be eliminated or damped to a point where velocities in it are not high enough to give fish a directional stimulus, then the downstream velocity of water leaving the fishway entrance will be the chief attraction to fish reaching this area, and there will be no delay in entering the fishway. This would be the case in the idealized condition shown in Plan B, Figure 3.3.

Smaller eddies, boils, and upwellings are also undesirable in this area, and for certain rock conditions, expensive additional excavation might be required downstream of the fish entrance, to a depth equal to or exceeding that of the spillway bucket, to eliminate such conditions. This gives a larger volume of water in which to dissipate the energy of the turbulent flow and reduces the tendency of the fish to become confused in direction at the particular time they should be heading directly into the fishway entrance. In general, however, the author has found that turbulence is a less serious hazard, especially where strong swimmers such as trout and salmon are involved, than the presence of a well-defined eddy with its mass of water circulating with appreciable velocity in a direction that does not encourage the fish to enter the fishway.

One general feature of fishway entrances located adjacent to spillways that has not been discussed in detail so far is the provision of attraction water or auxiliary water (see Figure 3.5, Section B-B). Its use is based on the concept that fish migrating upstream are attracted by velocities in a downstream direction of such magnitude that they are able to swim against these velocities and thus further their progress upstream. The normal flow through the fishway answers this requirement, but this flow is usually very small compared to the total volume of spillway flow. In addition, assuming that one has been successful in leading fish to the general area immediately below the fish entrance and adjacent to the spillway, it is necessary for these fish to detect the actual flow out of the fishway as quickly as possible. To make the flow out of the fishway noticeable at the greatest possible distance from the fish entrance, it must be speeded up to a velocity as high as possible without actually reaching the point where it might discourage or prevent entry, particularly of the weaker fish. The most generally accepted minimum standard for this velocity for salmon is 4 ft/sec. Considerable variation from this standard is possible, of course, without greatly affecting efficiency, particularly where other conditions such as turbulence might reduce its effectiveness. It is probable that flows of less than this velocity will in general result in less efficiency of attraction, while flows of higher velocity can increase efficiency up to the point where the swimming ability of the entering fish cannot cope with the velocity over the required distance. It is doubtful if 8 ft/sec may be safely exceeded even for the strongest fish, and velocities approaching this value should be maintained for only a short distance at the entrance of the fishway.

As long as there is flow through the fishway, practically any velocity can be achieved at the entrance by merely restricting the opening into the fishway. But this opening cannot be made so small that the fish have difficulty finding it or, in the extreme case, so small that they refuse to enter. It is probable that with a small opening, fish would have difficulty finding the entrance even with a high velocity through it, as the area over which the high velocity extends is also important. In other words, the quantity of flow at an entrance is important in addition to the velocity. A size of opening has to be selected, therefore, that is as large as it is economically and practically feasible to supply with water to meet the desired velocity condition. In practice, this opening is dependent to a certain extent on the quantity of auxiliary water, if any, which may be supplied to the fishway upstream of the entrance to meet the velocity requirements in the fishway when the downstream baffles are drowned out at high tailwater stages.

Referring once more to Figure 3.5. Section B-B shows an automatic telescopic weir on the main entrance to a fishway adjacent to a spillway, and also shows how the lower baffles of the fishway are drowned out at high tailwater levels. If it is specified that the minimum velocity in the drowned-out section of the fishway shall be 2 ft/sec, then a certain quantity of water will need to be added to the fishway flow here to meet this minimum velocity requirement during periods of high tailwater. In many cases this quantity of water will be adequate to maintain the desired velocity of 4 ft/sec over the telescopic or otherwise adjustable entrance weir.

If possible, the width of the entrance opening should be the full width of the fishway, and it might even be desirable to have a further auxiliary entrance on the side of the fishway next to the picketed lead as shown on the plan in Figure 3.5. This auxiliary side entrance is particularly useful if there is no picketed lead. The depth of water over the adjustable weir at the fishway entrance will vary according to the amount of attraction water used. For a simple low dam where auxiliary water is not used, the depth over the entrance weir will be approximately the same as the depth over the interior baffles of a weir type fishway, if the entrance is the same width as the fishway. This is based on a head of approximately 12 in. on the fishway weirs and a velocity of about 4 ft/sec over the entrance weir. In such cases it would be preferable to decrease the entrance width in order to increase the depth over the entrance weir. As the amount of auxiliary water increases, the depth over the entrance weir can be increased. For large installations where it is hoped to attract fish out of extremely turbulent waters below spillways, a minimum depth of 4 ft has been specified and velocities in excess of 4 ft/sec, as noted above, are found to be desirable.

To briefly review the foregoing, then, the desirable velocity at fishway entrances adjacent to spillways has been stated to be more than 4 ft/sec but less than 8 ft/sec for Pacific salmon. Weaker swimmers such as Atlantic Coast alewives and American shad would require lower velocities. The preferable width of the entrance has been stated to be equal to the width of the fishway, although for small installations some relaxation of the latter requirement might be permissible. For fishways of the vertical slot or orifice type, it probably would not be desirable to have an entrance the full width of the fishway, particularly if no auxiliary water were available. In this case the entrance would be shaped to resemble the orifice in the baffle rather than a weir. A preferable depth of the entrance has been set at 4 ft or more, but some relaxation might also be necessary in this requirement at a small installation on a low dam or under other special circumstances. Some comments have been made on the positioning of spillway fish entrances, but no specific comment has been made on the direction of the flow out of the entrance, although it has been taken for granted in the examples discussed that this flow is generally in the same direction as the spillway flow out of which we are attempting to attract the fish.

3.6 FISH ENTRANCES — POWERHOUSE (POWERHOUSE COLLECTION SYSTEMS)

The simplest fish entrances at powerhouses merely have the fishway emptying into a tailrace channel either adjacent to the powerhouse or at some distance downstream. This type of installation is fairly common although it will be seen by reference to Figure 3.4, Section C, that the farther downstream the entrance is located from the face of the powerhouse, the more difficult it will be for the fish to find. After they have reached the velocity barrier in the draft tube, they would have to return downstream to find the entrance to the fishway. When the fish facilities were designed for the first of the Columbia River dams at Bonneville, another problem was recognized. At this dam the powerhouse was expected to be extremely long, its 10 generating units of over 50,000 kW each extending for about 800 ft across part of the river. For this reason fish entrances at either end, even though they were immediately adjacent to the powerhouse, could not be expected to effectively attract fish that approached in the center of the powerhouse and remained in the outflow from the draft tubes at or near this point. A condition similar to that previously described for a wide spillway could be expected; however, in this case the powerhouse structure itself offered a convenient base in which to incorporate the solution. The flow after leaving the draft tubes is normally at much lower velocity and far less turbulent than in a spillway bucket; this made it possible to devise a flume with multiple entrances that could be attached to the downstream face of the powerhouse which the fish could enter at a number of locations across the breadth of the tailrace channel. This system designed for Bonneville Dam became known as a powerhouse collection system, and modified versions of it have been used at all the Columbia River dams built subsequently, as well as at many other dams elsewhere.

Between these two extremes of the simple entrance on the one hand, and the Bonneville type collection system on the other, there have been many powerhouse entrance systems devised. They all have a general resemblance in that the designers

**Figure 3.7   The entrance to the fishway in Pitlochry Dam on the River Tummel
in Scotland. The steel rack guides fish from the tailrace into the fish
entrance.**

have attempted to utilize the flow from the powerhouse as attraction water. The
fishway in the dam on the River Tummel at Pitlochry in Scotland is an example. Here
the entrance to the one fishway serving the dam is located in such a manner that the
flow from the powerhouse assists in attracting fish to the vicinity of the fish entrance,
and away from the spillway area. Figure 3.7 shows the fishway entrance, which is on
the right bank adjacent to the powerhouse but a short distance downstream. The
tailrace is screened by a rack with steel bars spaced sufficiently close to prevent adult
salmon from entering. This rack is placed at an angle across the tailrace so that it
leads upstream to the fishway entrance. The spillway extends across the dam from
the powerhouse to the left bank, and is not serviced by a separate fishway. It is
assumed that sufficient storage is available on the river system upstream so that when
the salmon are migrating, spills can generally be avoided or at the most limited to
short periods. This method of screening the tailrace channel to lead fish to the fish
entrance has been frequently used in Europe but is not common in North America.
Possibly the reason is that its success calls for a great deal more flow control than is
common in North America, plus a low debris problem. A distinct disadvantage is the
head loss on the turbines caused by the rack itself.

A number of existing fishways at powerhouses have entrances either immediately adjacent to or downstream of the face of the powerhouse, and are apparently giving satisfactory service. The author believes, however, that the collection gallery with an entrance or entrances immediately over the draft tubes is the most efficient method of attracting and collecting fish, at least for Atlantic and Pacific salmon. An opportunity presented itself to make a direct comparison of the two types of entrances with Atlantic salmon at the Tobique plant on the Tobique River in Canada. Here a collection gallery had been installed with an entrance over the center of one draft tube, but in addition an entrance was installed at the bank adjacent to the powerhouse and still another entrance some distance downstream. All these were in operation after the dam went into use but it was found that virtually all the fish used the entrances over the draft tube and adjacent to the powerhouse. Consequently, the downstream entrance has since been abandoned.

While it is desirable to use the collection gallery with multiple entrances along the downstream face of the powerhouse, it is still necessary, particularly on long powerhouses, to provide one larger major entrance at one or both ends of the powerhouse. This entrance should be equal in width to the main fishway, and it can be connected to the collection gallery by means of a so-called junction pool. Figure 3.8 shows schematically a typical junction pool for a large powerhouse collection system. Auxiliary water is required to meet the same standards set out for the fishway entrance adjacent to the spillway. In fact, at many locations, the powerhouse will lie adjacent to the spillway, and another fish entrance at the point where the two structures meet can serve one end of the spillway as well as the offshore end of the powerhouse. However, in this case the fish entering would be led through the powerhouse collection gallery and up the fishway on the bank at the opposite end of the powerhouse.

Perhaps it would be appropriate at this stage to examine in more detail an entire powerhouse collection system for one of these large installations, while some simpler, more economical installations will be discussed later in this text.

It has been noted that the width of the major entrance at the shore end of the powerhouse should equal the width of the fishway itself. This criterion has probably evolved from the reasoning that if the entire powerhouse happens to be shut down during the peak of upstream migration, the entire run may have to enter the fishway through this major entrance. Therefore the entrance should have the same width as the fishway itself in order to accommodate the maximum numbers present. For large installations where it is necessary to attract fish from wide, turbulent areas, it has been specified that the depth of the water at the major entrance should be a minimum of 4 ft, the same as specified earlier for an entrance adjacent to the spillway. This depth has to be controlled by telescopic weirs, of course, in order to keep velocities at the entrance within the range of 4 to 8 ft/sec. The quantity of auxiliary water required to meet these conditions and the degree of adjustment required in the telescopic weirs depend largely on the gross fluctuation in tailwater level expected during the period of migration. Careful consideration of timing of the run in relation to probable river flows and power plant operation is necessary to avoid unforeseen

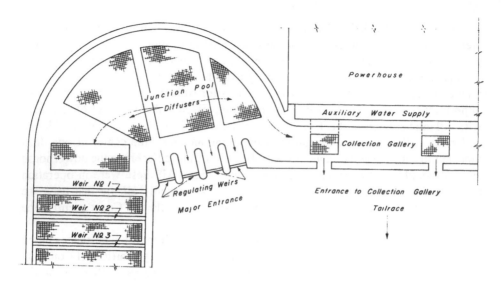

Figure 3.8   Sketch plan of a typical fishway entrance adjacent to a powerhouse collection gallery with junction pool serving both.

combinations of water levels that will be detrimental to fish passage and also to avoid unnecessary expense in providing for conditions that may occur only infrequently.

For example, although migration may have already commenced at the time of a spring freshet, the peak of the freshet may always occur when only a few of the early arrivals have presented themselves for passage. In this case it might not be worth the extra cost of providing for all the criteria set out for operation of the entrance at maximum efficiency during the highest river stages. Study of the history of the timing of migration in relation to the expected discharges should enable the designer to choose a maximum river stage up to which all entrance flows and design criteria must be met. At flows above this stage, auxiliary water flows will remain constant, and specified velocities will decrease as the adjustable weirs reach their limit of operation and become further submerged.

Peak discharges may only occur over comparatively short periods of time (say, at the most a few days and often only during one day), so that it would not be economical to provide for fully efficient operation of the entrance facilities to take care of such a transient condition. The same solution may be desirable in this case, that is, to set a maximum river stage up to which all entrance flow and design criteria must be met, but beyond which operating efficiency is allowed to gradually decrease.

The foregoing hypotheses assume that it is early in the run, and although passing the fish presenting themselves is still desired, somewhat less efficiency than normal is acceptable. This would not be desirable if the run were near its peak, when the fish with the maximum potential for reproducing themselves are normally present. In rarer cases it might be safe to consider making fish passage not only less efficient, but impossible, by limiting the time of operation of the facilities to a period less than the normal migration period, if by so doing worthwhile economies in operation can be effected. Such a limitation, which would probably result in delaying the early arrivals for a certain length of time, would necessarily depend on a biological assessment of the effect of such a delay on the segment of the run affected, and should not be considered unless there is a firm assurance from responsible biologists that it is safe.

While the main criteria for entrance design have been stipulated, it is sometimes desirable to specify additional features to give added assurance of protection to the fish. For example at McNary Dam on the Columbia River, Von Gunten et al. (1956) reported that the entrance weirs were specified to function automatically and to be sensitive to 0.1 ft in tailwater elevation fluctuation. They also report that a minimum discharge of 1000 cfs was specified for any main entrance. The latter is by no means a standard specification, but was established to meet the particular conditions expected at McNary. The specification for sensitivity of entrance weirs can be more generally applied, however. It has become fairly uniform on the Columbia River dams, and was cited by the Department of Fisheries, Canada, and International Pacific Salmon Fisheries Commission (1955) as a preferred standard for the most modern and efficient fishway it was possible to design at that time.

The multiple entrances to the collection gallery at the Bonneville powerhouse were weirs capable of adjustment to meet the specified velocity conditions at tailwater levels expected during migration. However, more recent experience has indicated that submerged ports might be more effective for attracting Columbia River chinook salmon into the collection gallery. Consequently, at McNary Dam, which was built some years later, the entrances were equipped with gates capable of operation as overflow weirs or as submerged orifices. Considerable flexibility was also incorporated at McNary in that a minimum of six entrances were provided for each of the 14 units in the powerhouse, half of which were blocked off with stoplogs, with the remainder gated as described. This permitted one half (actually 44) of the entrances to be operated as submerged orifices or weirs in the locations where most success was obtained in attracting fish as flow conditions changed. Because of the length of the powerhouse (1403 ft) it was feared that if fish were permitted to enter the long collection gallery at the main entrance situated at the junction of the spillway and powerhouse, they might be tempted to leave the fishway through one of the many entrances en route, so that a second gallery was provided behind the first. This second gallery, known as the *closed channel*, is connected only to the main entrance at the junction of the spillway and powerhouse, and to the junction pool at the opposite end of the powerhouse, so that fish can only proceed straight through it and into the main Oregon shore fishway.

The same velocity criteria that govern the design of the major fishway entrance are usually stipulated for the multiple entrances to the collection gallery, that is, a minimum of 4 ft/sec, and preferably capable of regulation up to 8 ft/sec. On the other hand there has been considerable variation in the size and shape of openings used for these entrances. Because of the large number of openings, and the resulting large quantities of auxiliary water needed to meet the velocity specified, it is necessary to keep their size to the minimum that will permit entry by the fish. The Department of Fisheries, Canada, and International Pacific Salmon Fisheries Commission (1955) suggest that the size of the openings be variable from 6 to 12 ft^2. This would give flows through each port a minimum of 24 and a maximum of 96 cfs. The McNary gates are 10 ft wide, and the flow through the ports is maintained at 60 cfs. This gives a depth of flow of 1.6 ft and a velocity slightly less than 4 ft/sec when operated as a weir. When operated as a submerged orifice, the opening is 1.5 ft deep and only 8 ft wide, and is submerged 2 ft beneath the tailwater level. A horizontal slot such as this is probably better suited structurally to installation in the segmental gate required at each entrance to allow for variations in tailwater. Also, it is often required to set some arbitrary limit on the quantity of water available for each opening. Spreading the opening horizontally, rather than in the form of a narrow vertical slot, reduces the distance between openings and uses the limited quantity of water to best advantage. It is deduced that these two reasons are the main ones for making the submerged orifice entrance in the shape of a horizontal rather than a vertical slot. It may be that in the future, after comparative tests are available, a shape nearer to square, or perhaps a vertical slot, will be proven to be more effective, but there is no experience to indicate the need for this change at present.

It is necessary, of course, to add auxiliary water to any collection gallery with multiple entrances where the capacity of the entrances exceeds the flow in the gallery. This can be accomplished as shown in Figure 3.9 by use of a second conduit parallel to the collection gallery, with diffusers spaced as needed along the gallery floor. The velocity criterion most generally used for all diffusion gratings where auxiliary water is added through the floor of fishway systems is 0.25 ft/sec through the gross area of the grating. Where water is added through the walls, this velocity has been increased to 0.50 ft/sec, but the reason for this 100% increase is not clear.

The diffuser system shown schematically in Figure 3.9 has been used frequently and is accepted as the most economical and practical method of adding auxiliary water. Ideally, the water would be diffused into the gallery at a uniform velocity over the whole area of the grating. This can be achieved only with an elaborate system of pipes and orifices, however, which is costly and troublesome. As a result a more practical arrangement is used, such as the stepped bottom shown, and any resulting lack of uniformity in velocity is compensated for by adhering to a conservatively low overall velocity standard. In the collection gallery it is probable that a group of diffusers will be needed for each set of entrance gates over each turbine. The water escaping through each gate has to be replaced to maintain the velocity in the collection gallery at or near 2 ft/sec, which is the accepted standard velocity for ensuring continuous migration of fish through open channels or sections of fishway where the baffles are submerged by high tailwater levels.

The methods of adding auxiliary water will be described in more detail in a following section. One other comment is appropriate here, however. With long powerhouses a consid-

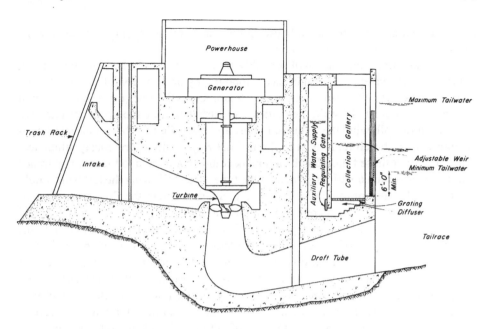

Figure 3.9 **Cross section through powerhouse showing collection gallery, auxiliary water-supply conduit, and diffuser.**

erable drop in water surface along the face of the powerhouse has often been found, particularly if the powerhouse is placed parallel to the flow in the river. If this slope is greater than the water-surface slope in the collection gallery (which is not very large if the velocity is maintained near the recommended 2 ft/sec), then it would be impossible to maintain the required head of 1 ft on the entrance weirs or ports, and there could be no uniformity of gate opening or entrance velocities across the face of the powerhouse. In such cases this problem has been overcome by placing adjustable baffles at intervals along the collection gallery walls, so that it is possible to duplicate in the collection gallery the surface slope of the water outside the powerhouse at all river stages. Care must be taken that these baffles do not create a concentrated drop in water surface inside the gallery, which might deter the ascent of fish. Because of the ready possibility that fish will leave the gallery through one of the many entrances if deterred at any point, a maximum concentrated drop of only 6 in. has been used as a standard.

It will be noted that in Figure 3.9 a depth of 6 ft below minimum tailwater has been specified as the lower operating limit of the control gates or weirs at the fish entrances. The collection gallery is also shown to have the same minimum depth. Besides providing for an adequate depth in the gallery at minimum tailwater level, this arrangement allows the adjustable weir to be lowered sufficiently to give adequate depth over the weir at minimum tailwater level. This 6-ft minimum dimension is of course ideal, and under some circumstances such as at very small installations it can be reduced.

The major entrance at the shore end of the powerhouse has been discussed, as have the collection gallery and its multiple entrances. The remaining major entrance is that at the junction of the powerhouse and spillway if they adjoin, or at the offshore end of the powerhouse if they do not.

This last major entrance will probably be the one that attracts the fewest fish unless there are exceptional conditions of topography or flow. By referring to Figure 3.10 it can be seen that fish migrating upstream along the margins of the river will naturally be separated by the velocities and turbulence of the spillway into two groups. Those following the right bank must all be handled by the major entrance A in Figure 3.10; therefore, it must be the full fishway width as previously stipulated. The majority of those following the left bank could be attracted to the major entrance B only under certain conditions. Most of the time those following the left bank would be attracted by the entire tailrace flow and to a lesser extent the flow from the left-bank end of the spillway; therefore, multiple entrances to the collection gallery and the third major entrance C are provided. The size of entrance C is a matter for judgment on the part of the designer. It should exceed the size of the collection gallery entrances and should probably be as wide as the gallery itself, but it need not be as wide as the main fishways or the other major entrances. If there is a chance that the powerhouse could be shut down partially or completely while the spillway continued to spill substantial volumes of water and fish migration was in progress, it would probably pay to increase the size of entrance C to something approaching the other major entrances. If not, dimensions closer to the minimum noted above would probably be satisfactory.

Some of the principles involved in the design of entrances to fishways at dams have been set forth in the foregoing paragraphs illustrated by use of hypothetical examples. At this stage we refer to some actual installations to demonstrate how these principles have been applied, and also to show how it has occasionally been necessary, because of economics and the physical factors involved, to use discretion in applying some of these principles. This, of course, does not mean that in those cases where some of the principles have not been followed exactly, the facilities will fail. Lacking proof to the contrary, it is assumed that at the existing facilities adult fish passage is successful to the degree desired in approving the design. In other words, some efficiency in operation must be sacrificed in those cases where the best design standards cannot be met.

Figure 3.11 shows general plans of two existing mainstream dams on the Columbia River, the Dalles and the McNary. Fish entrances are shown as well as the fishways. These entrances have been designed following most of the principles set out previously, because the value of the salmon runs protected has amply justified the most elaborate and efficient facilities. Some minor exceptions may be noted, however, even under these circumstances. An example that appears to be an exception but is not, is the major entrance to the powerhouse fishway on the left bank at McNary Dam, which faces substantially at right angles to the direction of the river flow rather than downstream as recommended in the text. This arrangement was followed because the tailrace channel is located in a bay or indentation in the river bank, so that the flow out of the tailrace is in a direction almost parallel to the face of the powerhouse. The principles have therefore been followed even though it appears that they may not have been.

Figure 3.12 shows plans of smaller hydroelectric installations, which incorporate facilities for passage of both Atlantic salmon and Pacific salmon. For economic

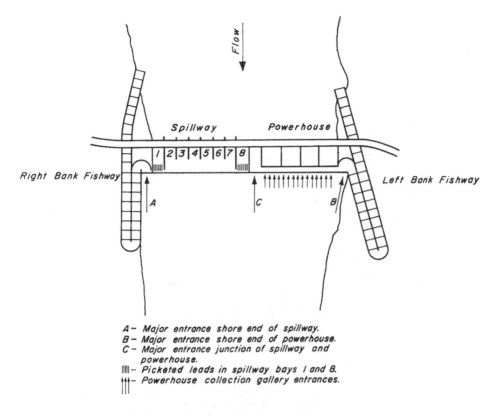

A - Major entrance shore end of spillway.
B - Major entrance shore end of powerhouse.
C - Major entrance junction of spillway and powerhouse.
|||| - Picketed leads in spillway bays 1 and 8.
††† - Powerhouse collection gallery entrances.

Figure 3.10 Schematic plan of typical hydroelectric dam showing fish entrances needed to meet normal requirements.

reasons these installations are less elaborate than those on the Columbia River; most of them have been designed to pass only a few thousand salmon upstream annually. The Pitlochry installation illustrates the Scottish practice of using a single fishway with the entrance adjacent to the tailrace. The tailrace is screened to divert adult salmon to the fish entrance, and spillway flow is infrequent enough during adult migrations that an additional (spillway) entrance is not warranted. Atlantic salmon utilize this submerged orifice fishway, which has an 18-in. head difference between pools.

The Seton Creek fishway shown is referred to later in connection with the use of hydraulic models. The dam includes facilities to pass adult Pacific salmon of the sockeye, pink, chinook, and coho species, and both resident and anadromous trout on their upstream migration. Here one fishway was considered adequate for the size of the runs involved and the width of the river. The entrance was placed next to the siphon spillway, which is always the first part of the spillway to come into operation. Attraction water was provided alongside the fish entrance by means of a small sluice with its own gate controls. This arrangement cost less than a standard diffusion system and was considered adequate for attraction purposes on the basis of the hydraulic model studies described elsewhere in this text.

The departures from the more elaborate plans used on the Columbia River are evident in these plans, which are presented to illustrate some typical simplifications

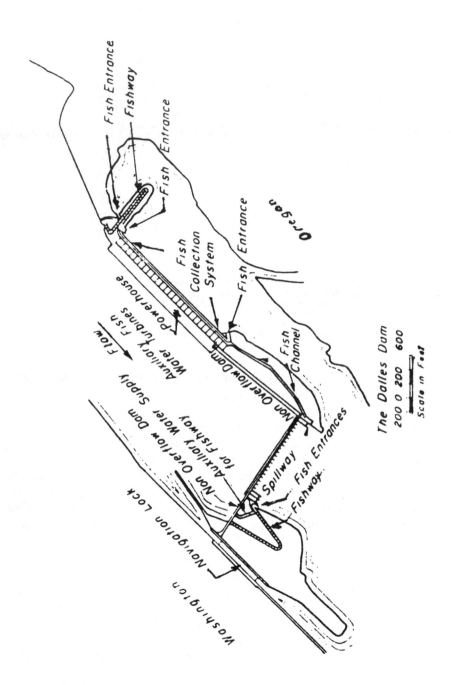

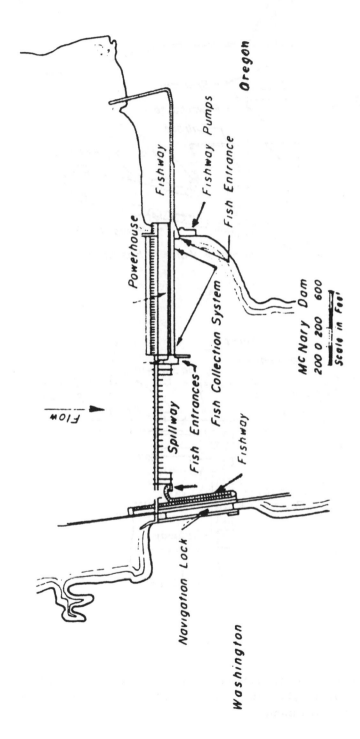

Figure 3.11 Plans of The Dalles Dam (above) and McNary Dam (below), showing the main features of the dams and fishways.

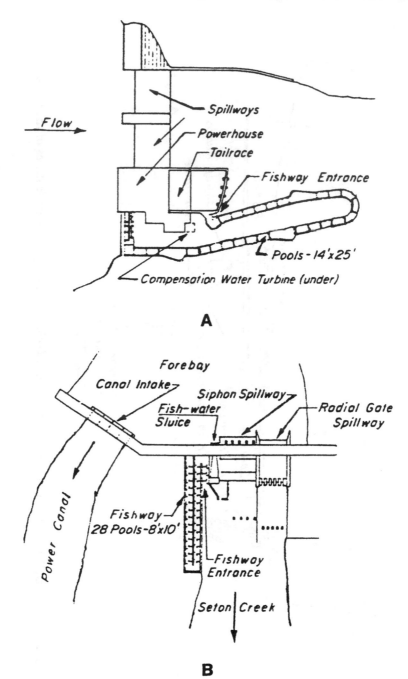

Figure 3.12 Plans showing fish-collection and -passage facilities at (A) Pitlochry Dam in Scotland and (B) Seton Creek Dam in Western Canada.

that can be used to advantage under certain conditions. They also seem to emphasize that facilities for fish passage must be designed to meet the particular conditions expected at any one dam, and therefore any general set of principles or rules that one can devise must always be applied with discretion.

3.7 AUXILIARY (OR ATTRACTION) WATER

As explained earlier, additional water over and above the normal fishway flows is needed for two reasons, first, to extend the area of intensity of velocity of outflow from the fish entrances to attract more fish, and second, to provide velocities in fish transportation channels of sufficient magnitude to encourage the migrating fish to keep moving in the required upstream direction.

The facilities for introducing auxiliary water through the floor or walls of the fishway have been described, but not the methods of getting it to this point. These methods are extremely important, as the introduction of highly aerated or turbulent water not only will be detrimental to the operation of the fishway, but will discourage fish from entering or passing through, possibly resulting in delay or damage to the fish.

The best way to ensure that there is no excessive turbulence and aeration in the auxiliary water supplied to the fishway is to specify that it be furnished from a low pressure system and that no air be permitted to enter the system. Water can be obtained to meet these requirements in several different ways. It can be taken directly from the forebay and passed through an energy dissipator to bring it to the desired head before passing it into the fishway. It can be taken from the forebay and passed through a special turbine designed to operate under a head somewhat less than the total head on the dam, and then to deliver it at the required low head to the fishway. Or it can be pumped directly from the tailrace to the diffusion system at the required head.

An economic study might be required to determine which of these methods is best for any particular situation, provided that the methods are not ruled out by obvious physical limitations. For example, in a case where the flow of a river is completely regulated, it would probably not be economical to take auxiliary water from the forebay through an energy dissipator, whereas in a run of the river plant where excess water is available most of the time, this would probably be the best method. If this method is not economical, then a study could be done to compare the economics of pumping attraction water directly from tailwater, against the economics of supplying it from the forebay through a special turbine to the auxiliary system. It should be remembered that to ensure the safety of the runs of fish against failure of the pumps or generators, standby pump capacity or a standby gravity supply directly from the forebay is required.

The most satisfactory delivery head is that just needed to induce the required flow through the diffusion chamber into the particular section of fishway or collection gallery where the water is required. If it is greater than required, it will create too much turbulence in the fishway or transportation channel. A maximum of about 6 ft of delivery head should be adequate for this purpose.

At McNary Dam, where some of the auxiliary water is taken from the forebay through a closed system, the energy is dissipated in a vertical expansion well in the wall of the fishway. Horizontal flow from the supply pipes impinges directly on a vertical steel-lined wall. The flow is then directed downward through an expanding section and into the diffusers in the fishway floor. Because the system is closed, all valves are subjected to heads varying up to almost full operating head on the dam, and because they could not be vented, satisfactory design was difficult to achieve. Considerable maintenance is expected to be necessary as a result of cavitation.

While these problems are not met in a low-head pumped supply from the tailrace, there are numerous other hydraulic problems inherent in the need to maintain all operating standards as the tailwater fluctuates. These are, however, fairly straight-forward hydraulic problems, which need not be elaborated in this text.

In auxiliary water systems there is only one other main problem besides those of conducting the auxiliary water and delivering it in the required condition and quantity to the fishways. That is the problem of preventing small downstream migrant fish from entering the system. This is a problem only to the extent that such migrants might be harmed by passing through the auxiliary water supply system. Once again referring to McNary Dam, it was apparently decided that potential injuries in the high-pressure system from the forebay were sufficient to justify screening the auxiliary water intakes with traveling mechanical screens. This type of installation will be described in more detail in the section on fish screens in Chapter 6. The low-pressure pump system supplying auxiliary water to one of the McNary fishways is, on the other hand, apparently not screened. It is therefore suggested that consideration should always be given to screening this auxiliary supply, particularly if the dam is high and the supply is drawn from the forebay. However, the method of energy dissipation and, in the case of pumping from the tailrace, the actual location of the pump intakes and type of pumps will have a large bearing on the final decision as to whether screens are used.

3.8 USE OF THE DENIL FISHWAY AT DAMS

In a previous section the reader was advised that it was desirable to make a preliminary selection of the type of fishway to be used as early as possible in the design stage. After that, in developing the various details of entrances and auxiliary water supplies it is assumed that the type selected was a pool and weir fishway (weir type) or a pool and orifice, or some combination basically involving stepped pools. However, it would be misleading to suggest that these were the only choices available to the designer. There is, of course, the Denil fishway. This involves, as previously described, a more intricate baffle design and probably a steeper gradient. It has been used extensively at dams on the East Coast of North America and throughout Western Europe. Provided that the critical criteria of limited fluctuation of headwater and tailwater have been met, it has functioned satisfactorily.

For example, the Denil fishways described by McGrath (1955) at Alvkarleby and Bergeforsen in Sweden were used to lead ascending fish to traps and holding ponds, where a stable flow was guaranteed as a result of the necessity to hold the level of the water steady in the ponds and traps. The Denil fishway is well suited functionally for this type of use, not only because of the complete control of flow available, but also because of the reduction in fishway maintenance possible by reason of the use of the controlled flow. Bedload and floating debris would be removed from such a system at some point upstream, and this would result in the elimination of the prime causes of excessive maintenance likely to be encountered in Denil fishways.

In cases where it is possible to use the Denil fishway at a dam, the same principles would apply in locating the entrance as previously described. The Denil type uses more water for its depth and width than any other type of fishway, and provided that the water is available from the forebay, this feature is a definite advantage in attracting fish to the entrance. If consideration is given to adding still more attraction water, this might be accomplished by using an entrance pool with the type of auxiliary water supply described, and having the Denil fishway lead upstream from the pool, although such an arrangement has never been tried to the author's knowledge.

The baffle design recommended by the Committee on Fish Passes (1942) for a 3-ft-wide Denil fishway called for the baffles to be placed on 2-ft centers (two thirds of the channel width) and sloped upstream at an angle of 45° to the floor. The floor slope was specified as not greater than 1:4. The clear opening in the center of the fishway was 1 ft 9 in., with the operating depth between 2 and 3 ft, which corresponded to flows of approximately 10 and 21 cfs. As noted earlier in the McGrath report, Furuskog revised these dimensions for the Hurting installation in Sweden to increase the volume of the fishway and of the flow, but in so doing felt it necessary to decrease the slope to 1:6. The Denil fishway at Dryden Dam, tested by the U.S. Fish and Wildlife Service and reported by Fulton et al. (1953) was based on this increased size and decreased slope. It had a width of 4 ft 3 in., and a length of about 30 ft. Ten U-shaped wooden baffles at 45° to the floor were spaced at 2 ft 10 in. on center. The clear opening between the baffles was 2 ft 6 in. according to the diagrams shown in the report. The best operating depth was stated to be 3 ft, corresponding to a flow of approximately 30 cfs.

In their review of Denil fishways, Katopodis and Rajaratnam (1983) state the following:

> Since the successful testing at Dryden Dam, the Denil fishway has been widely used in the Pacific and Atlantic coasts of North America. Decker (1967) reported that, in the state of Maine, Denil fishways up to 227 m in length (including resting pools) have been built and hydraulic heads of up to 15 m have been accommodated. Channel widths have varied from 60 to 120 cm and in all cases, the designs were geometrically similar to the one recommended by the Committee on Fish Passes (1942). Most fishways were built on a 16.67% slope, although a few were installed on a 12.5% slope. The vanes were sloped 3 vertical to 2 horizontal making the angle between them and the floor approximately 47 degrees and 49 degrees for the 16.67% and 12.5% slopes, respectively. Winn and Richkus (1972) evaluated the passage of alewife *(Alosa pseudoharengus)* through two Denils on the Annaquatucket River, Rhode Island. DiCarlo (1975) described several Denils installed in Massachusetts.

It is assumed that many if not all of these fishways were constructed at dams, and that they were generally satisfactory.

Larinier (1983) gives some guidelines for Denil fishways to be constructed in France for salmon and trout, based on extensive hydraulic model studies done by himself and A. Miralles, and his experience with installations of this type at dams.

He recommends a fishway 0.8 to 1.0 m wide for salmon, and 0.6 to 0.9 m wide for trout. This is for a type similar to that recommended by the Committee on Fish Passes. The gradient should ideally be about 16 to 20%. Clear width between baffles should be 0.583 times the overall width. Larinier also describes two other versions of the Denil, the "Fatou" and the "suractifs," which are similar to the steep pass version, and gives standards for these.

For the Committee on Fish Passes type he specifies a minimum depth of flow of 0.5 times the width, and an overall depth of fishway up to 2.2 times the width. This suggests a range of headwater fluctuation of about 1.5 times the width or a maximum of 1.5 m (6 ft).

Thus, it can be seen that Denil fishways are used extensively at dams, provided that the conditions are satisfactory. These conditions might be summarized as (1) a limited headwater and tailwater fluctuation and (2) control of slope and other factors so that the sustained swimming speed of the species to be accommodated is not exceeded in the fishway. If the entrance conditions described previously can be met, there is every reason to expect that this type of pass will be successful.

3.9 USE OF THE VERTICAL SLOT FISHWAY AT DAMS

The vertical slot baffle, described in Section 2.2, can be used in fishways at dams where applicable. It is particularly adaptable to low dams where variations in head- and tailwater levels are reasonably equal and in phase. Because of their use at natural obstructions where flows are uncontrolled and vary greatly, the designs developed to date are rugged and durable. When used at dams, they require little maintenance. Figure 3.13 shows the baffle plans used in the fishways at the Great Central Lake storage dam and at the Seton Creek diversion dam, both in Western Canada. Runs of more than 100,000 Pacific salmon, of the sockeye, coho, and chinook species have passed through the Great Central fishway annually with no apparent difficulty. The Seton Creek fishway has the same capacity. As noted earlier, the baffle can be constructed of steel plate or timber to the same or almost the same dimensions as the reinforced concrete baffle shown. The usual practice for all types of materials is to support the baffle by means of a beam across the fishway at the top and another across the floor or incorporated in the floor if a slab is used. The main baffle column near the center of the fishway (really a vertical beam) then spans from top to bottom between these two, as does the small one that juts out from the wall. The rest of the baffle wall can then be designed as a two-way slab connected to the main column and the fishway wall on the sides, and the top beam and floor beam at the top and bottom as shown for the Great Central fishway.

The Seton Creek design shown in Figure 3.13 is slightly different, however. Here the baffles have been completely cantilevered out of the fishway walls on each side. While no detailed cost comparison has been made between these two methods of design, it is believed they would cost about the same. The first method, using the cross beam at the top, facilitates the placement of gratings over the fishway if these are considered necessary. The usual arrangement for fishways at dams is to have a protective fence around the structure and no gratings, which not only reduces the cost, but permits those responsible for operation, as well as the public, to easily view the fishway in operation.

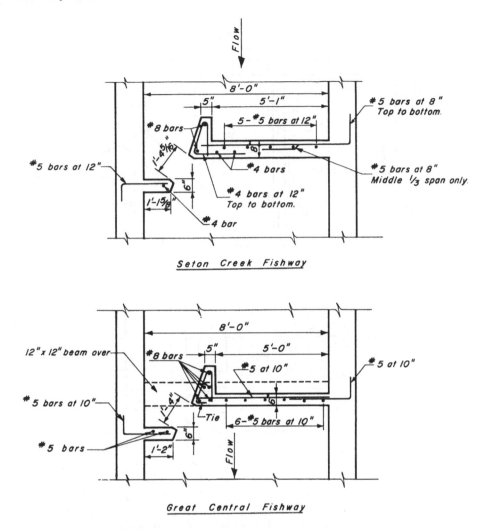

Figure 3.13 Baffle plans of two vertical slot fishways used at dams in Western Canada.

The same comments on slot width and other dimensions governing the hydraulic operation of the vertical slot fishway apply here, as outlined previously for this type at natural obstructions.

The vertical slot fishway has also been used successfully at dams in Eastern Canada and more recently in France. The version used at Bergerac on the Dordogne River in France shown in Figure 3.14 is based on the full Hell's Gate design shown in Figure 2.1. The maximum head to be surmounted by the fishway is 3.5 m (11 ft). The fishway has 11 baffles, and with the head loss at the entrance, this amounts to an average drop per baffle of less than 0.3 m (1.0 ft) at the most extreme low-water stage. At higher river stages the difference between headwater and tailwater decreases to 2.5 m (8 ft) as will be seen from Figure 3.14.

The fishway is 6.0 m (19.6 ft) wide and the baffles are spaced at 4.5 m (14.75 ft) on center. The slot width is 55 cm (21 in.), and other dimensions are as shown in the

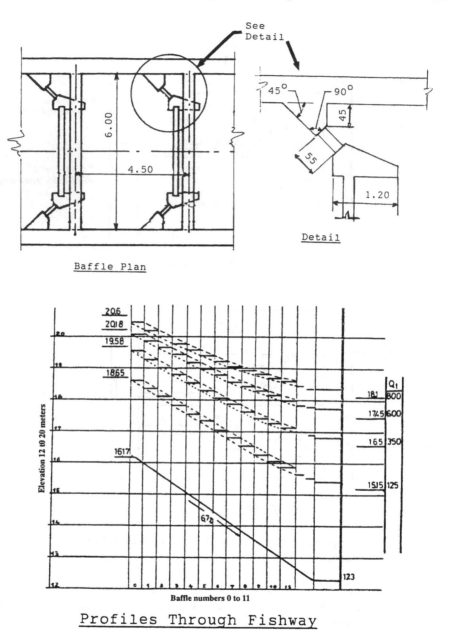

Figure 3.14 **Vertical slot fishway at Bergerac on the Dordogne River in France. The dimensions are in meters and centimeters. (From Larinier, M. and D. Trivellato, 1987.)**

figure. According to Larinier (1983, 1988), the shad they were planning to pass (allice shad) exhibited similar characteristics to American shad in that they preferred to migrate in schools near the surface of the water and became trapped in square corners. Therefore, the space behind the small stubwall was filled in and the overall head per baffle was restricted more than usual for European fishways. The fishway was modeled at a scale of 1:22, and an auxiliary water conduit was included at the side as a result, to improve entrance conditions.

This fishway has had runs of about 8000 shad per year since its completion, and it is hoped that 20,000 shad will be accommodated annually in the future.

A unique adaptation of the vertical slot fishway was designed and tested by the Forest Service of the U.S. Department of Agriculture for use at a natural obstruction in Alaska. It was never built, however, but its value as a type of fishway for use at dams is sufficient to warrant description here. It has a spiral shape and is designed to fit in a circular pipe with a diameter of 20 ft. The fishway itself is 5 ft in width, and the slot is 10 in. as shown in Figure 3.15. The pools are each one sixth of the circle in area (excluding the 10-ft circle in the middle), and the slope drops 1 ft per baffle. Provision was made to adjust the height of the sill in each slot to meet special flow conditions. Since some allowance must be made for this sill, a maximum variation in headwater of about 4.5 ft is all that could be accommodated with this particular design.

The design was modeled in a hydraulics laboratory at a scale of 1:3 and proved satisfactory, chiefly because the jets from each of the slots were directed into the wall and cushioning area of the next baffle, as shown in the figure. For larger headwater fluctuations, this fishway could be adapted to a larger-diameter pipe section, but care should be taken to model such an adaptation since the pool configuration would be changed with the change in pipe diameter.

3.10 USE OF THE WEIR TYPE FISHWAY AT DAMS

For the weir type fishway, the structural design of the baffle is straightforward. At the many existing installations, weirs have been constructed of most of the common building materials and designed to span between the walls or to be cantilevered from the floor. The shape of the weir crests, however, and the size, shape, and placement of any orifices in the weirs are important features that require more detailed consideration.

Some of the different shapes of weir crests that have been used at various fishway installations are rounded, beveled (on either the upstream or the downstream corner), and broad flat-crested. The latter shape has usually been accomplished by superimposing a flat plate on the crest of the weir so that it projects beyond the weir face in either the upstream or the downstream direction. Many publications warn that sharp crests should be avoided because the fish might be injured on them. It is, of course, unlikely that a very sharp weir crest would prove practicable to maintain. The different shapes of weir crests do alter the flow characteristics in a pool, so that if one shape has been proved to have an advantage in encouraging ascent of a particular species it should continue to be used.

When McNary Dam on the Columbia River was placed in operation, it was found that less flow than the ideal produced a transverse wave in this long fishway which reached an amplitude of as much as 8 ft. The result was an unstable flow that retarded fish migration. After a series of tests, it was found that beveling the top of the baffle as shown in Figure 3.16 remedied the problem.

Notches in the weir crest are commonly used to confine most or all of the flow to the notched portion, thus decreasing the water required for the fishway and

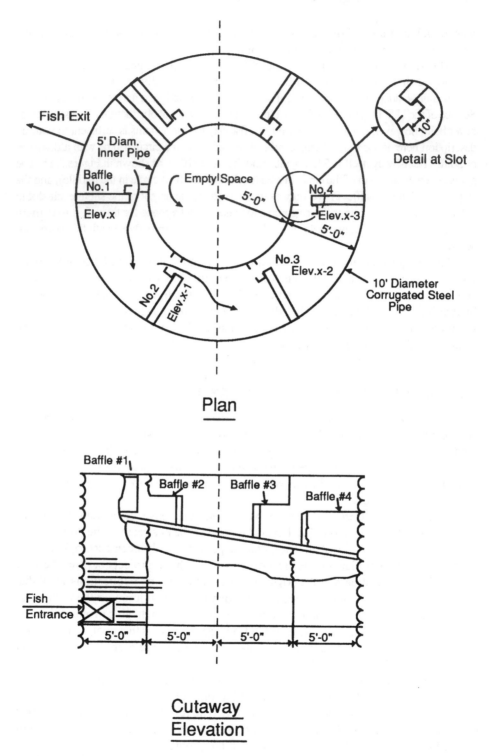

Figure 3.15 Helical vertical slot fishway designed by the U.S. Forest Service for use at Wolf Creek.

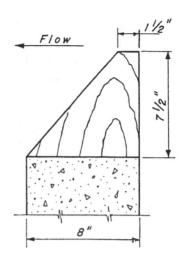

Figure 3.16 Beveled weir crest developed by the U.S. Army Corps of Engineers for use at the Dalles Dam to damp the tendency to oscillate found in long fishways at McNary Dam.

increasing the energy dissipation in the pools per unit of volume. Many versions have been constructed, some with notches in the center, and some with notches on one side or the other. In the latter case the notches have often been staggered from side to side on successive baffles, but staggering the notches is not generally recommended as this provides a tortuous path for the fish to ascend.

Another type of notched weir crest has been developed for passage of alewives on the Atlantic Coast of North America. Conrad and Jansen (1987) report as follows:

> We have found that sloping apron weirs are necessary for effective passage of this species. The aprons are approximately $2^1/_2$ ft in length and are placed on a slope of 3 horizontal to 2 vertical. Sharp edged weirs cause plunging flow, whereas sloping aprons direct the flow to the normal swimming depth of the fish and the slope of the flow is readily navigable. Alewives move in schools and pass through the baffle notches several at a time, side by side.

This baffle is illustrated in Figure 3.17. The same principle could apply to other species that migrate in schools.

Since construction of the Dalles Dam, further laboratory studies by the U.S. Army Corps of Engineers have resulted in a baffle design that is even better for reducing the tendency of the water surface to oscillate. In this design the weir has a high center section with short wings or stubwalls projecting upstream from the weir face at each end of the high center section as shown in Figure 3.18. The high center section comprises roughly the middle third of the weir, so that water flows over the outer third on each side in normal fashion. This pattern was subsequently used at Ice Harbor Dam as well as at other fishways in the Columbia basin.

Many fishways have used orifices to carry the entire flow between successive pools, as this has many hydraulic advantages over the weir and overflow type. One must always keep in mind the preference of the fish, however. If they will not ascend through an orifice readily, orifices passing all the flow should be avoided. In this case, orifices have been used to advantage in combination with weirs. The Columbia River fishways have orifices combined with weirs, and there are many other examples. Flow conditions over weirs are quite sensitive to changes in pool level, and the presence of

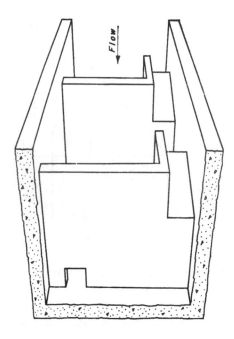

Figure 3.17 A type of weir used to pass alewives in fishways on the Atlantic Coast of North America.

submerged orifices through the weirs tends to stabilize the flows to a certain extent. In addition they provide the fish with a choice of passage upstream either through the orifice or over the weir. This is a valuable feature where a fishway is required to pass a number of different fish species, some of which might not pass over weirs as readily as most salmon and trout. Even for salmon, there is room for doubt as to whether a submerged orifice is not as effective as a weir. The use of weirs has always been justified on the basis that a series of weirs with water flowing over them most closely resembles the natural environment of salmon and trout. It should be remembered that the depth of water over the weir is probably far less than that which the fish are accustomed to swimming in, so that the use of weirs may be less of an advantage in this respect than is claimed. In addition, many fishways with submerged orifices and no weir flow between pools have successfully passed even the most active salmon. It is considered therefore that the pool and orifice fishway might be used more frequently in years to come.

As noted, the baffle with a submerged orifice through which all the flow passes has the advantage of being more flexible than a weir or combined weir and orifice when headwater changes are a problem, particularly if such changes are rapid. Flow through an orifice varies roughly in proportion to the square root of the head whereas flow over a weir varies in proportion to the 3/2 power of the head, other factors being equal. Bonnyman (1958) describes how this can be used to advantage in fishways, because variations in headpond level have very little effect on flow in the fishway. If all orifices throughout the fishway are uniform in size, the total head between headpond and tailwater tends to be distributed evenly among the baffles. If the headwater drops, say 2 ft, then this decrease in total head is distributed among all baffles, making very little difference in head on each individual baffle. Flows therefore remain almost constant in the fishway during small headpond fluctuations. Care must be taken to set the upstream orifice low enough to draw water at minimum headpond level, however, and it follows that the water in the upper pools of the fishway must be deeper than the

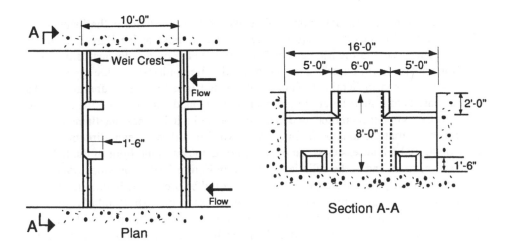

Figure 3.18 Details of the Ice Harbor Baffle developed by the U.S. Army Corps of Engineers.

orifices. For reservoirs with wide limits of draw-down, this problem has been overcome by having gated ports at the reservoir leading to conduits that serve the upper and several successive pools of the fishway. Usually it is sufficient to have the ports and conduits leading to alternate pools or possibly only to every third or fourth pool, as the design and economics of the situation permit. In operation, this type of installation would have the top port to the upper pool open at full reservoir. As the level drops to a point where this port ceases to be submerged, the next lowest port is opened, connecting to the third pool from the top of the fishway or a lower one as the case may be. As the headpond level drops, the reduced total head on the fishway is distributed among a smaller number of pools. The most important design considerations are to ensure that the fishway operates at all reservoir levels and that no conditions are set up that are detrimental to fish passage. The latter precaution has in mind the possible use of long, submerged penstocks to carry fish from the fishway to the forebay at low reservoir levels. Long, unlighted sections of fish-transportation channels such as these penstocks have generally been avoided in fishway design. Research at Bonneville Dam by Long (1959) indicates that these fears may not be entirely warranted, however. Long found that test fish (chiefly steelhead) actually *ascended* a darkened fishway faster than a well-lighted one. His tests were not extensive, however, and for one thing did not indicate whether fish would hesitate to enter a darkened fishway from a brightly lighted one. Delay on entry could be as important a contributor to total delay as slowness in ascent. Until more is understood about fish behavior, it is considered advisable to make any totally enclosed passages as short as possible and allow as much light as possible to enter. Artificial lighting could be considered where total darkness is unavoidable.

Flow control to the fishways by penstocks from the forebay to the upper pools can be designed to be automatically regulated by float controls. Such a system has the advantage over a system of tilting weirs such as that used at McNary Dam in that the regulation does not need to be nearly as sensitive.

Like weir crests, orifices between pools have been constructed in a number of different shapes. Rectangular orifice openings are easier to construct and are probably more common, particularly on smaller fishways. The orifices in the fishways at Bonneville and McNary Dams are rectangular, while those at Ice Harbor Dam are square. Both these and the McNary orifices are beveled on the downstream face.

Bonnyman (1958) describes a type of inclined cylindrical or pipe type orifice that has been used in Scotland. Thoroughly tested in a hydraulic laboratory, this orifice consists of a short length of cylinder with an inside diameter of 2 ft 3 in. The length of the cylinder recommended is about one and a half times the diameter. It is recommended that the axis be inclined downward in a downstream direction at an angle of 20°, which necessitates the provision of a pocket in the floor of the fishway at the downstream end of the cylinder as shown in Figure 3.19. Bonnyman points out that both a longer pipe and steeper inclination would be preferable hydraulically, but would not be as satisfactory in providing easy access to the fish. Similarly, the size of the cylinder diameter is governed largely by the requirement to provide ease of access by fish. Bonnyman implies that diameter less than 2 ft could be so confining as to restrict movement of fish and that larger openings would create too much turbulence in a fishway of this size. Bonnyman points out that the upstream edges of the inclined cylindrical orifice are rounded, with the result that the coefficient of discharge is about 0.9. This results in a flow of 39 cfs at 1.5 ft of head per baffle. The average velocity in the pipe is therefore almost 10 ft/sec.

The operating head of 1.5 ft per baffle is higher than the standard 1.0-ft head used for Pacific salmon, but heads in the range of 1.5 to 2.0 ft are commonly used for Atlantic salmon.

There is no doubt that orifices of other shapes and sizes than those described could be used with varying degrees of success. Increasing the total area would be particularly beneficial **if there is not danger of creating turbulence** in the pools because of the increased discharge. This relationship of pool size to discharge will be discussed further in the next section on fishway capacity.

3.11 FISHWAY CAPACITY

To the owner of a dam, one of the more important elements in the design of fishways will be the matter of fishway capacity. The width, depth, and length of the fishway determine its total volume or fish-carrying capacity, as well as its cost, so that capacity is related to cost, although perhaps not in direct proportion.

There will, of course, be a minimum acceptable size of fishway, which will be determined partly by hydraulic requirements and partly by the behavior of the fish in confined spaces. This minimum size will be elaborated on later in this section. For the minimum-size fishway the total length is governed only by the operating head on the dam. For the purposes of the following discussion, the head or height of the dam will be assumed to be constant.

We will assume for the present that the adult migrants for which we are designing the fishway are sufficiently numerous to require one larger than the minimum size. Let us examine the known methods of calculating capacity.

In Chapter 2, the method recorded by Jackson for Hell's Gate is described. Similarly, a method used by the author for fishways at natural obstructions is referred to in the same chapter. It should be noted that both of these methods have been applied

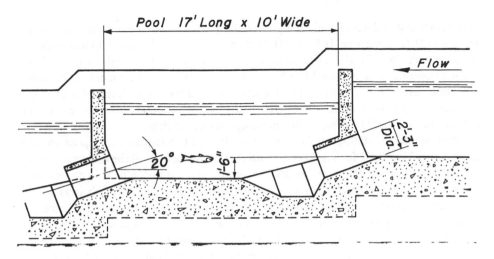

Figure 3.19 **Sectional elevation of a submerged orifice fishway with short cylindrical orifices which has been used in Scottish fishways. (From Bonnyman, G.A., 1958. *Hydro Electric Engineering,* Vol. 1, Blackie and Son, London, pp. 1126-1155. With permission.)**

to natural obstructions, and for reasons given at the beginning of that chapter, fishways at dams require a more conservative approach. A Dam is an artificial barrier on the migration route of the fish, and every effort must be made to remove any possible cause of delay in migration associated with it. We have described some of the elaborate precautions taken to reduce delay at the entrance of the fishway. Similar precautions are necessary to avoid delay in the fishway itself as a result of overcrowding of the fish.

The probability that fish can be delayed by overcrowding in a fishway is generally accepted by biologists. The Department of Fisheries, Canada, and International Pacific Salmon Fisheries Commission (1955) state that it is known from work on the Fraser River that crowding of fish in a limited area reduces freedom of movement and slows their passage.

They go on to state that if fishway capacity is exceeded, the number of fish passing could be expected to diminish rather than increase. Lander (1959) states that a fishway that is too small may hinder the spawning migration of anadromous fish and thus reduce survival of the next generation.

Elling and Raymond (1956) report on work at the Bonneville Engineering Research Facility, and the difficulties involved in setting up experiments to yield useful information are clearly evident. In commenting on this initial work by Elling and Raymond, Lander states that for one of the experiments at least, his conclusion was that movement through the test fishway was hindered by crowding.

If we accept that fish passage can be hindered by crowding, the next logical step is to try and define exactly the point at which harmful crowding occurs. This will be discussed later, but in the meantime we will examine some examples of practice in calculating the capacity of large fishways at dams.

While the writer is not aware of any complete written record of the method of verifying the capacity of the Bonneville Dam fishways, it is believed from discussions with some of those responsible that the following is approximately the method.

It is assumed that the maximum daily run to be passed through either fishway would be 100,000 fish and that 10%, or 10,000, of these would present themselves for passage in the hour in which peak rate of migration occurs. It is further assumed that the average rate of ascent would be 5 min per pool (with 1 ft of drop between pools). The third assumption was that 4 ft^3 of space was required for each fish if delay from crowding was to be avoided. On the basis of these assumptions it can be calculated that at a rate of 10,000 fish per hour, spending 5 min in each pool, the average number per pool would be $(10,000 \times 5)/60 = 833$ fish. If each fish requires 4 ft^3, pool volume must be $4(833) = 3332$ ft^3. The pool size used at the lower ends of the fishways, which would govern, were 30 ft in width, 16 ft in length, and 6 ft in depth. This gives a total volume of 2880 ft^3 which is close to the required volume. Both fishways widen upstream to widths of 38 and 42 ft, which gives more than the required volume per pool. Data accumulated at Bonneville Dam on the rate of migration through the fishways indicates that the lower pools of a long fishway became crowded before the pools farther up. This was not known at the time the Bonneville fishways were designed, which might account in part for the lack of provision for this eventuality.

It should be kept in mind that the foregoing assumptions and calculations were made before there was any dam in existence on the lower Columbia River, and in view of the lack of data such as actual fish counts, plus the lack of precedent for similar large-scale fish facilities, the assumptions and conclusions appear to have been remarkably good. It might be well to point out here that Bonneville Dam was constructed in the 1930s and the calculations of fishway capacity were the first ever attempted, to the author's knowledge. In the case of the Hell's Gate fishways cited earlier in this text, Jackson had the benefit of the Bonneville experience 10 years earlier, which, although not published, was available to him.

Actual counts in the Bonnevile fishways have failed to reach the totals assumed, however, so that the maximum capacity has never been reached and we have no direct operational knowledge as to whether the capacity assumptions were correct. The assumed size of the maximum daily run could be expected to be in error by a large amount because of the lack of data, but the error was on the safe side and can be justified on the basis of providing the *safety factor* commonly accepted in engineering practice.

Justification of a safety factor is not difficult in the light of subsequent events on the Columbia River, and in view of the continuing lack of data on the basic biology of fish. In 1957, for example, the maximum daily count of all species at the Dalles Dam, 48 miles upstream from Bonneville was 27,683 fish, almost 8000 *more* than the maximum at Bonneville. There is no ready explanation for this difference. Even if the maximum daily migration at some point near Bonneville were known, therefore, it would still be essential to apply a safety factor to take care of the many unknowns affecting the rate of migration such as temperature changes, unexpected increases in escapement from the commercial fishery, etc.

The figures for the maximum daily and hourly runs are criteria for use at a particular site, and can be determined by field investigations such as tagging at or near the site. The criteria of volume required per fish and rate of migration through a fishway, however, are not site specific and could be applied at many different sites with adjustments for factors such as species, size, maturity, etc. In the example that follows, these criteria were altered slightly to compensate for a difference in species and location, and because further data became available in the interim.

The Department of Fisheries, Canada, and International Pacific Salmon Fisheries Commission (1955) based the design of the best-known adult-fish-passage facilities for dams proposed at that time on the Fraser River on the following assumptions. First, they stated that a total of 4 ft^3 of space was required for each adult sockeye salmon, on the basis that 2 ft^3 are required for resting and 2 ft^3 for moving. It was considered more effective to provide resting space in each pool throughout the fishway and not in separate large *resting pools,* as has been done on the Atlantic Coast and in Europe, since it is not known exactly where the fish will require rest.

The rate of ascent was revised in the light of some experimental work done on sockeye salmon in the Bonneville fishways. This work consisted of counts of fish entering a fishway, and simultaneous counts at various intervals along the fishway for a distance of 35 pools from the entrance. As a result it was possible to calculate the maximum percentage of the maximum day's run of fish occupying a given number of fishway pools as follows: 25.4% in 35 pools, 18.7% in the first 24 pools, 15.4% in the first 17 pools, and 10.7% in the first 13 pools.

The new criteria of space and maximum rate of ascent were then applied to the proposed dams on the Fraser (which were of the order of 100 ft high) as follows:

- The maximum day's run is 750,000 fish (sockeye salmon).
- The powerhouse (right-bank) fishway is to pass 90%, or 675,000 fish.
- 35 pools are required to accommodate 25.4%, or 171,450 fish.
- One pool is required to accommodate 4900 fish (average).
- The pool volume required is $4 \times 4,900 = 19,600$ ft^3. If the depth of the pools is set arbitrarily at 10 ft and the length is set at 16 ft, then the width of the pools (and the fishway) = $19,600/(10 \times 16) = 123$ ft.

Other widths and depths could undoubtedly have been selected, of course, as pool proportions are governed (within limits) by hydraulic, economic, and structural considerations for pools of this size, rather than by requirements of the fish. There is a limit to the variation, however, as pools too narrow and long could inhibit the free movement of fish because of the restricted weir length or restricted number of orifices. Also, depths greater than 10 ft are not desirable in most cases unless it can be proved that the species to be passed normally utilize depths greater than this. Minimum pool length is governed mainly by hydraulic requirements, and unless the fishway is very small, it would not be possible to reduce the pool length below 10 ft without making it impossible to properly dissipate the energy. With a fishway such as the one described, where the width required is much greater than the pool length, it is preferable to use a minimum pool length of at least 16 ft.

The experimental work on the Columbia River in the Bonneville fishways also indicated that as much as 14% of the maximum day's run entered the fishway in 1 h. Applying this to the Fraser River example quoted above, we would have a maximum hourly run of 105,000 fish. At the average figure of 4900 fish per pool, the fish would have to ascend at a rate of $105,000/(4900 \times 60) = 3.57$ min per pool. This rate of ascent is within a range that has been verified by observation of groups of fish on the Columbia River.

Bell (1984), as mentioned earlier, has developed the criteria further, giving a space requirement relating the size of fish to the volume of the pool. He uses 0.2 ft³ per pound of fish. This results in 1.4 ft³ for 7-lb sockeye or 4 ft³ for 20-lb chinook.

Elling and Raymond (1956) quote an example of a rate of ascent of 3.3 min per pool based on the median elapsed time of a group of fish ascending a six-pool fishway. They admit, however, that in practice one is concerned with determining the rate of passage of all the fish through a fishway, whereas the median they use for convenience is based on only the first half of the fish in the group. Since their tests were based on only a proportion of the group ranging from 61 to 92%, there was no way of checking the median against the actual mean. It would appear possible that there could be a considerable difference as a result of a few very slow fish. This would lead one to regard the rates of passage for all their tests, which ranged from 2 to 5.8 min per pool (median elapsed times in a six-pool fishway of 12 to 35 min), with caution. Elling and Raymond were, however, not concerned directly with determining the rate of passage but were attempting to determine fishway capacity, which involves the space requirement as well as the rate of passage.

As noted earlier, Lander (1959) concluded that one of the tests conducted by Elling and Raymond showed that movement was hindered by crowding. This particular test resulted in a maximum number of fish per pool of 239 in each of the first two pools. On the basis of the pool volumes given, this amounts to 2.5 ft³ per fish. The maximum number of fish in the fishway was 948, which amounts to 4 ft³ per fish for the total six pools.

It is possible to develop a very simple formula for determining fishway capacity, and by means of the formula to compare the various criteria outlined so far in this text for pool fishways with 1 ft of head between pools. The following symbols are suggested:

C = fishway capacity, in numbers of fish per hour
V = pool volume, in cubic feet
v = volume required per fish, in cubic feet
r = rate of ascent, in number of pools per minute

Thus V/v is the maximum number of fish a pool can accommodate, and $60r$ is the rate of ascent in number of pools per hour. Multiplying these, we have the following expression for the capacity of the fishway in fish per hour:

$$C = \frac{V}{v}(60r)$$

Transposing, we have

$$V = \frac{C(v)}{60(r)}$$

It is possible to solve this equation very quickly for pool volume if we know the value of C, which is equal to the maximum hourly run expected and can be determined by biological methods, and the value of the expression v/r. The latter combines the two criteria noted of volume per fish and rate of ascent. The following table gives the value of this expression for the examples quoted previously in this text:

	Rate of ascent $1/r$ (minutes per pool)	Volume per fish v (ft³)	v/r
Bonneville Dam (1930) — chinook salmon	5	4	20
Fraser River, Dept. of Fisheries, Canada, and IPSFC (1955)	3.57	4	14.2
Columbia River, Elling and Raymond (1956)	3.3–5.8	2.5–4	14–15
Bell (1984) — general criteria	2.5–4	0.2/lb	—
Sockeye — 7 lb	2.5–4	1.4	3.5–5.6
Chinook — 20 lb	2.5–4	4	10–16

It must be remembered that these values can only be used to determine the capacity of pool type fishways with 1 ft of head difference between pools. Values of v/r for smaller species with differing swimming ability are only subject to speculation until tests such as those by Elling and Raymond have established them. However, it will be noted that smaller fish, which are likely to require less volume per fish, are also likely to require more time per pool in the ascent, so that the value V/r may not change much for other sizes or species. It would be desirable, however, to have many more tests made to confirm and expand the data presented here. In the meantime, it might be well to include a warning here against extrapolation of the results shown unless conditions can be expected to be reasonably similar.

Reference has been made previously to the existence of a minimum pool size, which is likely to be determined by the reaction of the fish to the confinement imposed by the walls and baffles, and by the necessity to provide satisfactory energy dissipation to avoid excessive turbulence.

For the vertical slot fishway, a pool size of 3 ft wide, 4 ft long, and 2 ft deep is probably the absolute minimum for fish up to 2 lb. This fishway would have a slot width of 6 in. and a head difference between successive pools of less than 1 ft. The slot width might be increased if less head is desired between pools, but care must be taken to avoid excessive turbulence, a tendency that can be observed at the 6-in. width of slot with 1 ft of head. For fish over 2 lb, up to the largest salmon, the absolute minimum pool size for a vertical slot baffle would be 6 ft wide, 8 ft long, and 2 ft deep. The slot width should not be more than 12 in., with a corresponding head difference of 1 ft between successive pools. The tendency toward excessive turbulence will be noted with these dimensions, and an increase in pool size to 8 ft wide by 10 ft long is desirable if it can possibly be achieved. Slot width can be increased for reduced head between pools. The 6-ft-wide, 8-ft-long, and 2-ft-deep pool would accommodate 374 fish per hour of a size and swimming ability similar to fall chinook salmon. The 2-ft depth should be regarded as an absolute minimum to be reached only when water conditions are unusually low. An increase to 4 ft would, of course, double the capacity as well as giving the fish more freedom to select a desirable swimming depth and to keep in the shadows if they prefer.

For the orifice fishway, Bonnyman (1958) quotes the hydraulic requirement that the minimum pool length should be six times the orifice diameter and the minimum width should be four times this diameter. This is based on a 1.5-ft head difference between pools, and is intended to ensure adequate energy dissipation. For a 2-ft-diameter orifice this results in a pool 8 ft wide by 12 ft long. It is felt from experience with the vertical slot, and with combined weir and orifice baffles, that a somewhat smaller orifice could be accepted, probably in the range of an 18-in. diameter, to produce a minimum pool size of about 6 ft wide by 9 ft long. Depth requirements would then be probably a minimum of 4 ft to ensure submergence of the orifice at all times and prevent air entrainment as far as possible.

For a weir type fishway, a pool size of 4 ft wide, 8 ft long, and 6 ft deep has been used for experiments at the Bonneville laboratory, and it apparently imposed no restriction on the rate of ascent of even the largest salmon. Once more, this was for a head difference of 1 ft per pool, and the capacity of this pool works out to about 800 fish per hour. While under some circumstances a narrow fishway such as this might be justified by a significant cost saving, in many cases it would probably pay to make the width 6 or 8 ft for convenience of construction, if for no other reason. If it is considered desirable to incorporate an orifice in the weir, it then becomes necessary to make the minimum width at least 6 ft, with a pool length of 8 ft, and depth of 6 ft.

For Denil fishways, the choice of sizes is limited if one wishes to adjust the size of fishway to the expected size of migration. Various sources suggest that the maximum practicable size of Denil is about 1.2 m (3.94 ft) wide and 1.75 m (5.74 ft) deep. Beyond this size, it is generally accepted that adding another Denil fishway alongside the first is the only way to increase the capacity.

To get an idea of the capacity of one Denil, it might be advisable to examine some of the few tests that have been made. Zeimer (1962) reported successful installations of the steep pass version up to 27.4 m (90 ft) in length and set at slopes in the 25% range. These were at natural obstructions ranging up to 6.7 m (22 ft) in height. He estimated the capacity at 750 fish per hour for Pacific salmon. Thompson and Gauley (1964) describe experiments at John Day Dam on the Columbia River also with a steep pass version of the Denil 6.1 m (22 ft) long, which passed as many as 2520 fish per hour. It was assumed that this was set at a slope of about 20%, thus surmounting a height of only 1.2 m (4 ft).

As noted earlier, Bell recommends using 15 sec elapsed time for a salmonid to pass through a Denil "section," and provision of an equal time in resting pools between sections. This is equivalent to passing 14 fish averaging 50 cm (1.6 ft) long every 15 sec through a section 7 m (23 ft) long in a tail-to-nose configuration. This is equal to 336 fish per hour. If they ascended two abreast, it would be 672 fish per hour. Therefore, the figure of 750 fish per hour does not seem unreasonable for maximum capacity.

These figures could be applied only to a dam of about 2 m (6.5 ft) in height unless a series of sections were used, connected by resting pools. The resting pools should have a volume to conform to the standard of 0.2 ft³/lb of fish to hold a minimum of 28 fish of the size indicated. Suppose a 50-cm fish averaged 7 lb, then 40 ft³ minimum would be needed for the volume of the rest pools.

No experiments have been done on the length of time the fish might take to rest in the connecting pools, but it would be unwise to assume that it would be a uniform 15 sec, particularly as the total length of the fishway and number of sections increased. For small runs of fish it might be feasible to surmount high dams by the use of sections of Denil fishways 7 m (23 ft) in length with connecting rest pools of arbitrary dimensions, but on the whole it would seem that the current practice of limiting their use to dams of low head with limited headwater and tailwater variation would be the safest.

It should be noted here that model studies were made by McLeod and Nemenyi on Denils without any bottom vanes. The advantage of this would be in extending the range of headwater and tailwater variation under which the fishway would operate by simply extending the depth of the fishway. Two models were chosen as being good energy dissipators, but the work was never followed up, either in the laboratory or in the field. For anyone interested, this would seem to be a lucrative field for study and experiment.

3.12 FISHWAY EXITS

Because the fish entrance location is the most important phase of the design of the fishway, it should always be decided first. The location of the fishway from this point is then mainly a matter of economics and is usually left largely to the discretion of the dam designer. Sometimes this results in an exit location that is acceptable, but not always. There are several points that must be kept in mind when deciding whether an exit location is suitable.

First, there is a danger, which must be avoided, of having the fish leave the fishway in an area where they might be readily swept back downstream over the spillway, or through turbines, or into an intake used for some other purpose. The fishway exit should therefore be removed some distance from the spillway, particularly because the velocities immediately above spillways are usually high. Velocities near the entrances to turbines and other intakes are usually lower, with 4 ft/sec being the normal velocity through a turbine intake protected by a trash rack or coarse bar-screen. For this reason there is not the same necessity to have the exit as far removed from these latter intakes as there is from the spillway area. It is still desirable, however, to keep it as far as possible from all intakes and in velocities as low as possible. The plan of existing dam layouts shown on previous pages will illustrate the application of this principle.

There is one exception to this principle, which applies when the fishway must act as an egress route for young fish (or adults) migrating downstream. In some areas, and Scotland is a notable example, there is a possibility of having no spill at a time when this migration is taking place. In addition, it is possible that the turbine intake or intake for other purposes is screened to prevent entry of the small fish. The only means of egress is then down the fishway, and the fish exit for upstream migration then becomes the entrance for downstream migration. Some attempt must, therefore, be made to locate the fish exit at a point where the downstream migrants are likely to accumulate or at least where they are likely to pass in close proximity. If there is another source of attraction to the migrants, such as a turbine intake, which is screened, good screening practice would dictate the provision of a bypass near the screen to carry the fish away. This bypass could lead either to the fishway or directly to tailwater. Bypasses will be described in more detail in Chapter 6, Section 6.

Some mention has been made previously of flow control in the fishway, the provisions that are usually incorporated in the exit section. The means of regulating flow in an orifice fishway by gated ports was described. In addition it is possible to set these ports so that no gates are necessary, different ports being available for use by the fish as the surface of the headpond is lowered. Such an arrangement becomes more difficult as the potential fluctuation in the headpond increases.

For the normal weir type of fishway, the upstream end is constructed with a horizontal floor, and weirs in this section are adjustable so that the flow into the fishway can be controlled and the top pool levels maintained in steps as the headpond level changes. Control can be achieved by using stoplogs for the weirs in this section, or by a more elaborate system of gates or tilting weirs. If the fluctuation in the headpond is gradual, changing only a few feet over a month or a season, stoplogs adjusted manually may be adequate; they are certainly the least costly form of control. If, however, there are changes in pond level of a quarter of a foot or more each day, it might be desirable to install some form of automatic control. The tilting weirs and telescoping control gate installed at McNary Dam are a good example of a more elaborate design of this type of control. They are shown in Figures 3.20 and 3.21. The telescoping weir (344) at the exit actually controls the flow to the fishway. It is automatically regulated by a float control to give the desired flow over a range

Figure 3.20 The tilting weirs at the upstream end of the Oregon Shore fishway at McNary Dam. View looking upstream into the forebay.

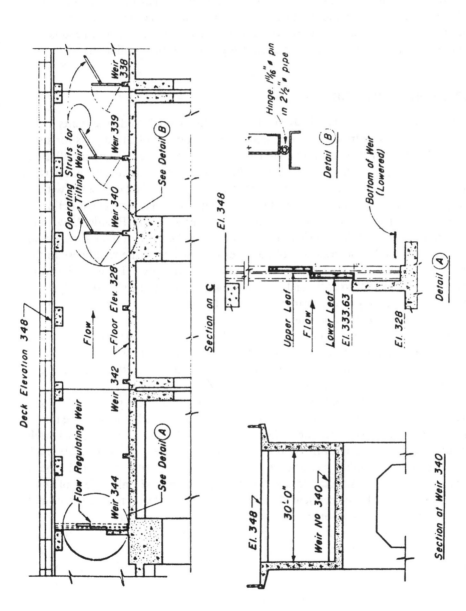

Figure 3.21 Section on center-line and details of control weirs at the head of the fishways in McNary Dam.

of 7 ft in headpond level. Below this are seven tilting weirs, which are also adjusted automatically to conform with changes in the headwater level. These weirs tilt about hinges as shown. They are operated by struts, which extend to the deck where a gear and motor is located. Because of the fact that sudden changes in volume of flow in the fishway can cause cessation of upstream migration, the automatic controls must be quite sensitive to changes in reservoir level. A sensitivity of 0.1 ft of head is now generally accepted as an optimum standard for this type of installation.

3.13 HYDRAULIC MODELS

Because of their comparatively small size, fishway pools and other parts of the fishway such as the entrance, reverse bends, etc., are almost ideal subjects for study by the use of natural scale hydraulic models. It is even possible in some installations to study a larger part of the project such as the fish entrance and an adjacent portion of the spillway, or even an entire dam and fishway at a fairly large scale without having to resort to a geometrically distorted scale. This permits construction of models that will conform closely to the requirements of the Froude Law, with gravitation being the chief physical force influencing the flow in the fishway components. If the model includes part of the river bed, however, the roughness in this portion will have to be adjusted to obtain dynamic similarity. This is achieved by constructing the model to scale and adding roughness where needed until dynamic similarity is evidenced by conformance of such features as the model water-surface profiles with the prototype profiles.

The Committee of the Hydraulics Division on Hydraulic Research (1942) gives typical scales for models of conduits (which would be somewhat similar to fishway models) ranging between 1:15 and 1:50. For models of rivers, this A.S.C.E. manual suggests a horizontal scale between 1:100 and 1:2000, and a vertical scale between 1:50 and 1:150.

These scales are probably more generally applicable to the type of model listed in the manual than to fishway models, which have tended to fall outside these ranges in many cases. This is understandable because the selection of a scale is largely determined by the facilities, time, and money available. The scale is normally made as large as possible within these limits. Models of short lengths of a pool type of fishway have been constructed to a scale as large as 1:6, and only a few of those familiar to the writer have been scaled smaller than 1:10. For Denil fishways the small size of the prototype has usually allowed them to be tested full size.

Model studies have enabled the various experimenters concerned to reproduce quickly a great many flow conditions in fishway pools by using different quantities of flow and different baffle designs. The phenomena known as *plunging* flow and *streaming* or *shooting* flow in a weir type fishway are shown diagrammatically in Figure 3.22 after Pretious et al. (1957). These phenomena have been studied both in the Bonneville Hydraulics Laboratory by the U.S. Army Corps of Engineers and in the University of British Columbia Hydraulics Laboratory by Pretious and Andrew, working for the International Pacific Salmon Fisheries Commission. None of the

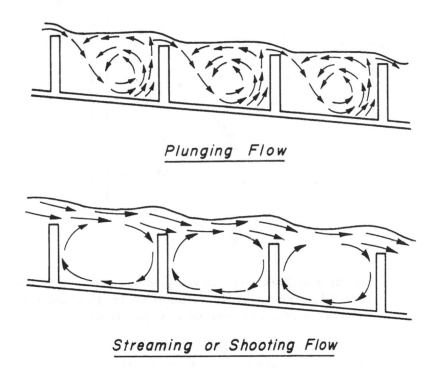

Plunging   Flow

Streaming   or   Shooting   Flow

Figure 3.22 Longitudinal section through weir type fishway showing the two conditions of plunging and streaming or shooting flow. (After Pretious, E.S., L.R. Kersey & G.P. Contractor, 1957.)

results have been published, but data on the tests have been made available on request. In general it has been found that stable plunging flow occurs when the head on the weirs is less than 1 ft, and stable streaming flow occurs when the head is 14 in. or more on the weir. There is some variation in these limits with changes in the shape of the weir crests, and with changes in the size and positioning of any orifices through the weirs. It can be seen from the diagram that when plunging flow exists, mixing in each pool is thorough, and complete energy dissipation is accomplished more uniformly throughout the pool. Under the streaming flow condition, the main part of the flow is confined chiefly to the surface layers in the pools, the lower layers being less subject to mixing, turbulence, and aeration. While most salmonids are able to ascend a fishway where stable streaming flow exists with a head on the weir of as much as 14 in., it is believed to be better to limit this head per weir to just under 12 in., or near the upper limit of stable plunging flow. At higher heads the flow in the models often proved to be unstable, changing at intervals from plunging to streaming, a condition reported by Elling and Raymond (1956) to seriously impede the progress of fish

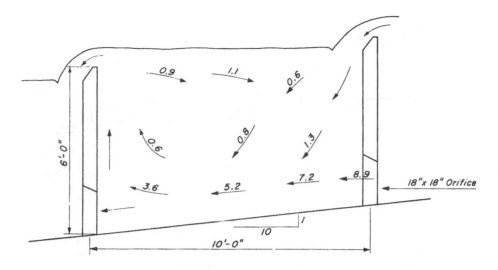

Figure 3.23 **Section through weir type fishway pool showing velocity measurements taken in model study. Velocities are feet per second (ft/sec), and dimensions are in terms of prototype quantities.**

through a fishway. Flows occurring at weir heads of about 8 in., on the other hand, in long, wide fishways, produced the surge noted previously in this chapter, which is also very undesirable. Figure 3.23 shows the velocity measurements that can be taken in a model fishway pool to determine the conditions that will confront the fish in their ascent. The velocity measurements can be taken with a Pitot tube or a midget current meter. The measurements are not precise, but give a good picture of the velocity pattern in the pool. Studies of these patterns will reveal any excessive velocities that might cause difficulty for the fish, and will also permit an estimate of the available resting area in the pool.

Vertical slot fishway pools have received intensive study by a number of investigating agencies. These studies have been necessitated by changes in the structural design of the baffles to suit physical conditions at a number of different sites. For example, in considering the type of fishway construction best suited to the Moricetown Falls site in Western Canada, it was decided that baffles of structural steel would be preferable. This choice was made because of the subzero weather conditions likely to exist during the proposed construction period, which would make the placing of concrete difficult. It was felt that model studies would be advisable in this case to see whether a single steel plate might safely be substituted for the 6-in.-thick concrete projection from the wall at each baffle used at Farwell Canyon (pilaster B, Figure 2.11). Some 31 tests were run on a scale model of a Moricetown pool, resulting in the plan shown earlier in Figure 2.2. At the conclusion of the tests, it was known that with the addition of a 6- to 12-in.-high sill on the floor at the bottom of the slot, this plan would provide satisfactory energy dissipation and fish-passage conditions. Some of the variables tested were the sill height, length of nose of the center column of the baffle, and width of the slot. It was noted that the presence of the sill at the bottom of the slot tended to pull the main thread of the jet from the slot over into the cushioning pool, where better mixing and more thorough energy dissipation took place. The hydraulic condition whereby the jet impinged directly onto the next center column downstream, or passed directly into the next slot downstream, was avoided, as this resulted in an acceleration of the water from pool to pool, until velocities became so high as to prevent passage of fish through the downstream slots. Other undesirable conditions that were eliminated by the model studies were surface instability or surge in the pools, large upwellings, extreme turbulence and aeration, etc.

Submerged-orifice fishways have not been used extensively in North America, and as a result, very few model studies have been made on them on this continent. An exception however, was a small amount of work done by the U.S. Army Corps of Engineers at Bonneville on regulating pool baffles to be used in the fishway at Ice Harbor Dam. Bonnyman, as noted earlier, refers to laboratory studies done in England that were invaluable in determining the shape, size, and inclination of the orifice, and the pool dimensions best suited for passage of Atlantic salmon. Undoubtedly the use of hydraulic models was well worthwhile in this instance.

Denil fishways have received extensive study in the hydraulics laboratory, commencing with the development tests undertaken by Denil himself in 1908. This was followed by the tests undertaken by McLeod and Nemenyi in 1940 and the Committee of Fish Passes in 1942. More recently they have been studied in Denmark by Lonnebjerg in 1980, in France by Larinier in 1983, and in Canada by Katopodis and Rajaratnam in 1983. This has resulted in a mass of data, from which a selection of a Denil for a particular situation should be possible. As noted earlier, Larinier has set standards for salmon and trout based on his studies. Standards for other species should be based on their swimming ability and preferences, as related to the experimental data on velocities, gradients, etc.

In addition to their use in the study of baffles and pools, hydraulic models have been used to study local conditions at particular points in fishways. For example, the conditions that will occur at right-angle bends or reverse turns have been studied by the use of large-scale segmental models of these parts of the fishways. From these studies, the undesirable upwellings that can occur at corners have been eliminated or reduced. Models of parts of the systems for supplying auxiliary water have been constructed to determine the best methods of eliminating aeration and cavitation. Virtually any part of a fishway system where a potential hydraulic problem exists that cannot be readily solved by theoretical considerations lends itself to a study by models.

The use of models to aid in the design of the spillways of dams is well recognized. Where fishway entrances are being incorporated adjacent to such spillways, they can be tested simultaneously. It is often the case, however, that only a portion of the entire length of the spillway is studied in a glass-walled flume, and in this case the fish entrances may have to be studied separately. It is possible to study fish entrance conditions by a model of the entrance and a short section of adjacent spillway, as was done for the Seton Creek Dam by Cooper and Boresky (1953). This model resulted in changes in the design of the dam involving deepening of the stilling basin by more than 5 ft, and the addition of deflector vanes. This is considered to be a good example of the use of a small economical model of a vital part of a dam and fishway that resulted in a great improvement in the hydraulic conditions for fish passage, as well as later benefits in decreased operating and maintenance problems. The Seton Creek Dam and fishway have been in operation now for many years and were proven to be satisfactory (Andrew and Geen, 1958).

In addition to model studies of pools and other parts of the fishway, and various parts of dams such as the fish entrance areas, modeling of the entire dam can be useful to those interested in fish conservation, particularly if the proposed construction period extends over one or more periods of migration. Construction problems will be discussed briefly in the next section, but it might be pointed out here that a model of the whole dam can very readily be altered to show river conditions at the various stages of construction. It is often necessary to construct cofferdams in such a manner as to result in increased velocities in the adjacent river to values that are beyond the swimming capabilities of the fish. These conditions can be foreseen by the use of a model of the dam as it will appear during the construction phase at the time of migration. Such models of the entire damsite are much larger than those discussed previously. Their actual size, of course, depends on the size of the river, the space and water supply available, and the scale chosen. For large rivers such as the Columbia, in order to keep the scale of the model large enough to permit accurate laboratory measurements, the model itself has to cover a considerable area and needs to be housed in a large building with a sizable water supply. Models of this type will require expert assistance from hydraulic engineers who specialize in this field. More information can be obtained on the subject of hydraulic models from many existing texts on hydraulics, in addition to the A.S.C.E. manual previously mentioned.

While models of complete dams and sections of river involve fairly high costs for facilities and staff, the smaller models, as shown in Figure 3.24 and described previously, are comparatively inexpensive. A typical series of tests of fishway pools, for example, might extend over a period of a month. These might be accomplished by only one engineer working alone, but at the most an engineer and a technician (to make alterations as needed) would be required for a month. Possibly two weeks of preparation would be required by the engineer to design the model and to adjust and calibrate it. The model itself could probably be constructed by one technician in two weeks, or at an equivalent cost if done in a large carpenter shop.

Both materials and labor would be higher if the model included a portion of the river bank or bottom molded in concrete. A further two weeks of engineering time would be required to analyze and write a report on the test results. The cost of laboratory space and equipment varies. Space in a public or semipublic institution such as a university hydraulics laboratory might be obtained at low cost. The cost of operating the large pumps necessary in model work will add to the total cost.

A typical small model test series might therefore cost as follows in 1986 U.S. dollars:

Engineering time: 2 man-months at $3,300	$ 6,600
Technician time: 1½ man months at $2,500	3,700
Materials (plywood, paint, hardware, etc)	1,200
Laboratory space and equipment; 1 month	2,800
Total	$14,300

**Figure 3.24 A small economical model of a dam and fishway in a hydraulics
laboratory.**

This total does not include overhead costs such as supervision and consulting
advice. The total could be lower for a simpler model or shorter series of tests, and
could run considerably higher for a more elaborate model including a part of river
bed.

In making this estimate, it has been assumed that an arrangement could
be made to have one's own staff members construct the model and conduct
the tests. Their time is therefore calculated at cost. If it is necessary to have
the model study done by contract with a commercial firm, the total amount
charged could be much more. The following is the estimated cost for a model
constructed under such conditions. The model includes a low dam 10 ft high
in a section of river 700 ft long and 300 ft wide, which carries flows of up to
20,000 cfs. The proposed scale is 1:36.

Construction of model and equipment to operate	$33,000
Program of tests and studies — two months at $18,000	36,000
Total	$69,000

While this model is more elaborate than the first example quoted and therefore
not strictly comparable, it can be seen that the direct charge for a commercial model
is considerably higher than for other models.

3.14 FISH PASSAGE DURING CONSTRUCTION

Normally in the design of any dam, careful consideration is given to the timing and sequence of the various stages of construction as they relate to seasonal changes in river flow. Dams are usually constructed in parts such as one third or one half at a time, each part in turn being surrounded by a temporary barrier known as a cofferdam, which prevents water from entering. It is often necessary to pump out the working area at considerable expense so that work on the dam can proceed in the dry, and soundness of the structure can be ensured. If the construction period lasts more than a year, any cofferdams in use during the period of seasonal freshets either must be constructed high and strong enough to hold out the floods and permit work to proceed through any flood period, or must be constructed to withstand overtopping, with the work within them stopped until the flood subsides and the area can be dewatered. The decisions as to where the cofferdams are placed, how high they are constructed, and what materials are used for them are governed by economics, availability of materials, and a number of engineering considerations.

An important additional consideration can be the necessity to assure passage of fish during the construction period. A common sequence of construction is to place the first cofferdam around all or a sizable portion of the spillway section of the dam. When the work in this area is completed, the cofferdam is removed, and a new one is placed around the remainder of the structure to be built, which could include the powerhouse at a hydroelectric dam. This last step involves diverting the river through the completed spillway section, and usually it is accomplished when the river is at its seasonal low-flow stage.

If fish are likely to be migrating upstream when the first cofferdam is in place, the velocities in the river due to the width constriction of the cofferdam must be examined critically to determine if fish can still ascend. This can be checked on a hydraulic model if one is available. If velocities are dangerously high, particularly along the banks (say, in excess of 15 ft/sec at spots or 12 ft/sec over an appreciable distance), it might be necessary to provide a temporary fishway along the outer edge of the cofferdam or on the opposite river bank. This fishway should meet the same standards as a permanent fishway, except that it can be constructed of less durable materials.

In some cases it may be sufficient, on the bank opposite the cofferdam, to provide a marginal path of lower velocities by the use of rock groins extending from the bank into the river. This method was used at Ice Harbor Dam on the Snake River as shown in Figure 3.25. In other cases the river channel may be wide enough to take the constriction of the cofferdam without increasing velocities seriously, and no extra provisions will be necessary to ensure passage at this stage.

If it were possible during this stage to raise the reservoir or headpond to its permanent level, there would be less of a problem because the permanent spillway fishway or fishways might be operated. However, this is rarely possible, as it is more often necessary to keep the head on the second-stage cofferdam as low as possible to avoid extra cost in its construction. A possible

Figure 3.25 **During construction of Ice Harbor Dam on the Snake River a number of rock groins were used along the right bank of the river to ease passage of salmon upstream through the narrow, high-velocity section caused by the first-stage cofferdam. One of these rock groins is shown in the foreground.**

compromise is to construct the spillway fishway in such a manner that it can be operated for this period at the extremely low forebay level in existence at the time. This might also involve considerable added expense, so that in practice a number of temporary expedients have been used such as trapping and trucking the fish, hoisting them over the spillway in a pail or bucket, or passing them through a shiplock. These methods, while only temporary and consequently affecting in most cases only one year's run of fish at each new dam, have proved for the most part unsatisfactory for Pacific salmon, and are not recommended. Even though temporary fishway facilities may prove more costly, they are recommended wherever possible to meet the problem.

3.15 FISHWAY COSTS

In estimating costs the same basic procedure is followed as for any other engineering structure. The accuracy of a preliminary estimate based on experience with similar structures depends entirely on the extent of the experience and the soundness of judgment of the estimator. As the project passes from the preliminary stage to the design stage, accuracy of the estimate depends less on judgment and more on meticulous quantity surveys of the project components as they materialize and become definite in size and composition. It is not proposed here to proceed any further than the preliminary stage, because it is not the purpose of this text to deal with such matters as quantity surveying and detailed estimating of construction costs, which are adequately covered in other textbooks. The unit prices and percentages suggested in the paragraphs that follow as a guide for making preliminary estimates were derived from exact costs determined from quantity surveys.

There are several indices available for determining the preliminary costs of fish facilities, all of which are based on experience in the U.S. and Canada. All have been converted to U.S. dollars. The safest method of applying costs derived from previously built fishways is the use of costs per unit of volume. Using this as a basis, it is found that the cost of a large fishway such as those at the dams on the Columbia River average about $37/ft^3 for the fishway structures alone, based on 1987 prices. This unit cost could be considered the top price for a heavy structure at a dam where very little advantage could be taken of the natural contours, and it was necessary to provide reinforced concrete piers or bents to support the structure at required elevations. Furthermore, this unit cost would provide for a floor slab and walls of reinforced concrete throughout. It could probably be reduced in those locations where it was possible to take advantage of natural contours and place the structure directly on solid rock, using the rock for the floor and walls of the fishway where suitable.

For fishways at dams, however, the cost of the fishway structure alone can only be considered as a base cost. Other costs will be added to it for the various appurtenances described in previous sections of this text such as the control weirs, auxiliary water supply, powerhouse collection system, extra entrances, etc. These latter costs will depend on the degree of automation desired, the quantities of attraction water considered necessary, and the physical size of the entrances and collection gallery. A summary of unit costs, intended for use in making preliminary estimates of fish facilities, follows on the next page. All costs quoted are on the basis of 1987 prices in U.S. dollars.

In addition to the initial cost of fish facilities, there will be operating and maintenance costs associated with them. These will be in addition to interest charges on the capital investment, and any provision for depreciation or replacement, both of which will be calculated on the same basis used for the dam to which the facilities are appurtenant.

On the basis of the few records available, it is estimated that total operation and maintenance charges on a large fishway installation should average between 1 and 2% of the capital cost. Slightly less than half of this will be for operation, and the balance for maintenance. Some additional allowance may be necessary for power lost in supplying water for the fishways and auxiliary water supply at times when no spill is necessary. The cost of this will depend directly on the head and the cost of the power produced, but in all probability it will not be more than a fraction of 1% of the capital cost annually, on the average.

Summary of Fishway Cost Criteria

1 Fishways at natural obstructions: Based on a number of vertical slot fishways averaging about 8 ft wide and about 10 ft deep (from floor to top of walls) with baffles at 10-ft intervals and surmounting obstructions up to 35 ft high; cost per cubic foot of volume encompassed by the structure ....$20–$40

2 Basic fishway structure at dams on large rivers: Based on a number of structures averaging about 30 ft wide by 10 ft deep (from floor to tops of walls) with baffles at 16-ft intervals and surmounting dams up to 100 ft high; cost per cubic foot of volume encompassed by the structure $37

3 Miscellaneous items appurtenant to the fishway, including control weirs at upstream end and at fish entrance: Based on fully automatic operation described in text ....29% of Item 2

4 Auxiliary water supply for fishways adjacent to spillways: Based on velocity criteria described in text; the amount can vary between 30 and 75% of Item 2, because it depends on the quantity of auxiliary water needed to meet thecalculations ....50% of Item 2

5 Auxiliary water supply for fishways at powerhouses with offshore entrance adjacent to spillway: Based on velocity criteria described in text. This also depends on the quantity of auxiliary water needed, which is larger than that quantity needed for the spillway fishway because of the multiple entrances to the collection gallery. The amount can vary between 50 and 130% of Item 2, and an average of 80% is recommended for preliminary calculations ....80% of Item 2

6 Powerhouse collection system: This varies almost directly with the length of powerhouse; estimate is per lineal foot of length of powerhouse for a dam on a large river having large runs of fish $12,628

7 Offshore entrance to powerhouse fishway: The amount varies from 6 to 10% of Item 2, and an average of 8% is recommended for preliminary calculations ....8% of Item 2

3.16 DESIGN CONSIDERATIONS FOR ANADROMOUS SPECIES WORLDWIDE

Throughout this text various criteria have been given for velocities, depths, and flows. These have been given in the units applicable to the particular situation being described, such as ft/sec for entrance velocities to fishways on the Columbia River. The following table is given to show the metric equivalent to these criteria for Pacific salmon:

Criteria	English	Metric
Desirable entrance velocity	4 ft/sec	1.2 m/s
Acceptable entrance velocity	4–8 ft/sec	1.2–2.4 m/s
Desirable depth of entrance	4 ft	1.2 m
Acceptable depth of entrance	1.6–4 ft	0.5–1.2 m
Velocity through diffusion gratings for auxiliary water	0.25–0.5 ft/sec	7.5–15 cm/s
Head difference between pools	1 ft	30.5 cm
Space required for fish	0.2 ft³/lb	0.00125 m³/lb
Maximum velocities in slots or over weirs	8 ft/sec	2.4 m/sec

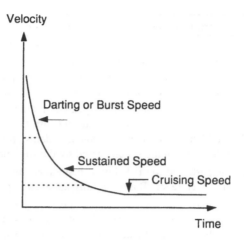

Figure 3.26 Schematic illustration of the swimming speeds of a particular species and size of fish at a particular temperature.

We now have to expand these as far as we can for other anadromous species.

The key factor in standard pool type fishway design is the drop or head per pool. What determines the head that a particular species can ascend? We know it is related to the swimming ability of the fish in question, but what relationship is there between the swimming speed of a particular species and the head it can comfortably surmount in a fishway?

Figure 3.26 is a generalized schematic diagram of the swimming ability of fish to which one might expect to assign values after considerable research. The three speeds will be recognized, as defined by Bell (1984) in an earlier chapter. Their definition is somewhat tenuous, however, and thus the exact speeds relating to each definition, and their boundaries, are very indistinct, even for species about which we have considerable data.

Figure 3.27 is an attempt to make these velocities more definite for certain of these species. Here the swimming speeds have been given from a number of sources, chiefly Bell (1984), in terms of the three speeds as defined earlier and shown in Figure 3.26. Some assumptions have had to be made since the definitions are not exact. The maximum darting speed is clear and appears on the figure as the speed that can be maintained for 1 s. The transition from darting speed to sustained speed is not so clear; for our purposes it is assumed to be the speed that a fish can maintain for 15 min.

The head drop between pools to be used in fishway design is obtained from Figure 3.28, which relates velocity to head. If we take for a standard the head between pools of 1 ft (0.3 m), which has been used for fishways for Pacific salmon for 50 years, we can determine from Figure 3.28 that it corresponds to a velocity of 8 ft/sec (2.4 m/sec) over a weir or through an orifice or slot. From Figure 3.27, we find that this corresponds to the speed the salmon can swim for a duration of 50 min.

Using this duration for the other species shown, we would expect that a fishway should have the following velocities and head between pools to have success equal to those for Pacific salmon:

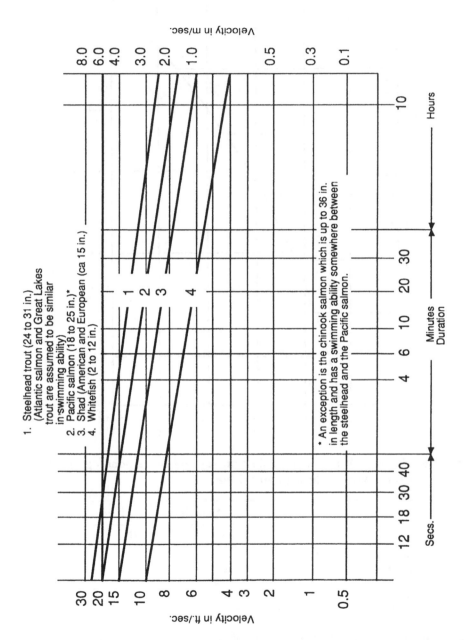

Figure 3.27 Swimming duration at velocities shown for species listed.

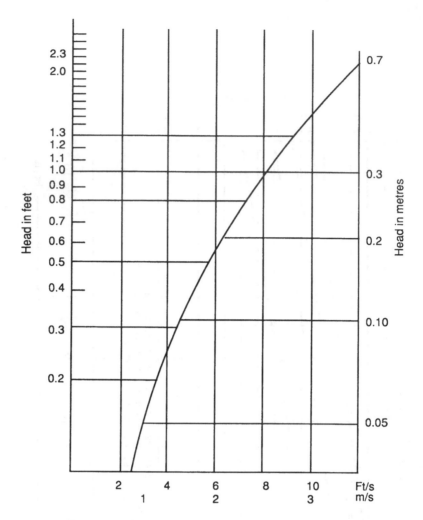

Figure 3.28 Head loss between pools of a fishway and corresponding velocities over a weir or through an orifice.

	Velocity		Head	
	ft/sec	m/sec	ft	cm
Atlantic salmon, sea trout, lake trout	12.0	3.4	2.3	70
Pacific salmon	8.0	2.6	1.0	30
Shad	5.8	1.9	0.5	15
Whitefish	4.0	1.3	0.25	8

It should be noted that the whitefish is not anadromous but diadromous, and is listed because it migrates upstream as an adult and meets the same problems as anadromous species.

It is interesting to compare these values to some listed by Conrad and Jansen (1983) and Larinier (1983), based on experience. See the following table.

Species	Assumed average length (cm)	Head loss per pool (cm)		
		Theoretical (from Bell, 1984)	Conrad and Jansen (1983)	Larinier (1983)
Atlantic salmon	60–77	70	61	30–60
Pacific salmon	60	30	—	—
Brook trout	30	—	30	30–45
Alewives	15	—	30	—
Shad	38	15	23	20–30
Smelt	20	—	15	—
Northern Pike	27	—	—	15–30
Whitefish	20–30	8	—	—

While there are some discrepancies, the general relationship between length of fish and allowable head loss per pool is evident in the above table. One would be wise to use the most conservative values shown and vary them only on the basis of experience with a particular species in a particular area. The alewives are a case of this variation, which is based on experience. They are a lively fish in spite of their small size, and tend to migrate in schools, as described previously.

For any species not mentioned, there is either insufficient, or no swimming-speed measurements, and no recorded experience. The only thing on which we can base fishway design is reference to the average length of fish shown in the above table and a guess as to whether they vary from those listed in swimming ability.

For the Denil fishway one cannot make similar general rules. As stated earlier, there are hundreds of combinations of baffle type, slope, depth, width, and length of Denil fishway in use in North America and Europe. It is difficult to rationalize these various combinations, so just one is offered here as a general guide. This is the one presented by Larinier (1983), as shown in Figure 3.29.

Although these specifications can be adopted with confidence for the species noted below, the reader would be well advised to consult the other sources noted throughout this text if additional study is warranted.

Atlantic Salmon[a]	
Width L (m)	Slope I (%)
0.8	20
0.9	17.5
1.0	16
1.2	13

Sea Trout[b]	
0.6	20
0.7	17
0.8	15
0.9	13

Note: Minimum depth of water in fishway is 0.4 · *L*.

[a]Preferred total width *(L)* of fishway is 0.8 to 1.0 m.

[b]Preferred width *(L)* is 0.6–0.9 m.

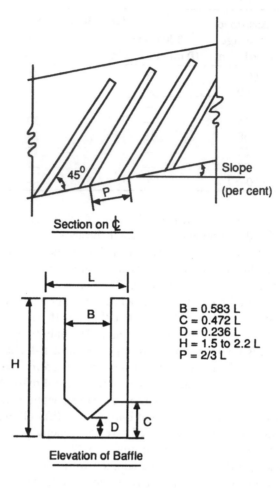

B = 0.583 L
C = 0.472 L
D = 0.236 L
H = 1.5 to 2.2 L
P = 2/3 L

Figure 3.29 Dimensions of Larinier's standard Denil fishway for salmon and trout. (From Larinier, M., 1983. *Bull. Fr. Piscic.*, July.)

It will be noted from Figure 3.29 that the angle of the vanes to the bottom is 45°, and the recommended maximum depth of the fishway is 1.5 to 2.2 · L **measured on the slope of the vanes.** One can readily see that by increasing the width of the fishway the slope must be decreased so that with the maximum width of 1.2 m the slope is 13%, which is very close to the slope of the vertical slot fishway shown previously. Therefore, it would seem that this is an upper limit for Denil fishways.

Now, how do these data apply to species other than salmon and trout? For fish that are weaker swimmers, one would expect that a lower slope of fishway with only a modest width should be used. In addition, it would be expected that it would be best to use a Denil of this type only for low dams with a minimum run of fishway to be ascended.

3.17 FISHWAYS FOR EELS

Eels are catadromous, breeding in the ocean and migrating into freshwater to mature. Fortunately they have characteristics that make them different from most catadromous species and therefore easier to provide a means of ascending past obstructions in rivers.

Eels are present in many of the streams and rivers in the world. As mature adults they spawn in at least three known areas; the Sargasso Sea in the Atlantic, east of Madagascar in the Indian Ocean, and west of Fiji in the South Pacific. The young are carried by ocean currents to the mouths of rivers and streams, where they then actively migrate inland to feed and mature in accessible lakes and streams. The whole process from larval stage to maturity can take up to 20 years. Some of the eels born in the area of the Sargasso Sea are carried by the Gulf Stream to the coast of North America, and others are carried across the Atlantic to Ireland, England, and most of the European countries with seacoasts. Some even enter the Mediterranean Sea and grow to maturity in Italy, Greece, and the countries of North Africa.

The eels born in the Indian Ocean are carried by currents to the southeast coast of Africa, entering the streams there. There are also eels present in the rivers of India. Those born in the South Pacific are carried to New Zealand and Australia, and some are known to be present in the Mekong River in Southeast Asia. Eels are also present in the rivers of China and Japan, but it is not known where these stocks originate.

When the eels reach the estuaries of rivers, they may be a year old and are known as glass eels because of their transparency. As they adapt to fresh water, they become pigmented and are known as elvers. It is in this elver stage, when they are from 5 to 15 cm in length, that they first encounter dams and other obstructions to their migration. As they proceed slowly up the river, they grow larger, and may be as much as 1 m in length and 10 years old, as in the St. Lawrence River in Canada when they reach the R.H. Saunders Dam and pass through the fishway.

Fortunately at the earliest stage the elvers are good swimmers. Tesch (1977) reports that eels 7 to 10 cm in length can swim at speeds of 0.6 to 0.9 m/s, while eels 10 to 15 cm long can swim at a speed of 1.5 m/s. These speeds can be maintained over a distance of 120 cm, which is in the range of their darting or burst speed. This is one of the criteria that can be used in the design of fishways for elvers.

Most of the early exploitation of the eel fishery was in Europe, where they are prized as a delicacy. Europe was also one of the first areas to exploit their water resources by construction of dams and weirs, and it was therefore natural that the problem of accessibility for eels was first recognized there and measures to combat the problem of dams were initially developed. These measures form the basis for what we know about eel passage at dams. All the other eel species, including the American eel, the shortfin eel of Australia, and the Japanese eel, have similar behavior patterns and capabilities, and what has been learned in Europe can be safely applied to other countries and species.

In addition to speed of swimming, the elvers have the capability to climb through brush and over grassy slopes, provided they are kept thoroughly wet. Eels have been passed upstream by means of flumes roughened to reduce the velocities, and by pipes full of wet straw and wood shavings. The flumes have had various systems of roughening, such as low baffles or cleats, placement of straw ropes on the bottom, or use of a mixture of brush, wood shavings, and in some cases, coarse gravel. In one case sacking was hung vertically over the face of a low dam, and as long as it was kept sufficiently moist, the elvers were able to ascend it.

All modern fishways for eels take advantage of this climbing ability, and when combined with their swimming ability the fishways are quite successful. A trough is normally used, which contains small branches sometimes mixed with wood shavings, and is usually placed at a slope of 12°, although some variation in slope has been used. Water is passed through in moderate amounts so as to thoroughly wet the branches and shavings but not to totally submerge them. The depth of water is enough to allow the elvers to swim around corners and reverse bends.

Modern eel passes have been improved still more in recent years by substituting nylon bristles and brushes and synthetic imitation branches for the natural branches and rope previously used in the flume. The reason for this is the greater durability of the synthetic material over the natural branches, which disintegrate in a few weeks or months, and thus have to be continually replaced.

Figure 3.30, from Jens et al. (1981) shows a cross section of a modern eel fishway on the Staustufe Zeltingen/Mosel in Germany. The inside dimensions of the flume are about 10 cm (4 in.) in depth by about 23 cm (9 in.) wide. The nylon bristles are mounted in the base vertically, and individual bunches of bristles are spaced about 1.5 cm apart. They are about 9 cm (3½ in.) long and the depth of water flow is maintained at about 5 cm (2 in.). The figure also shows how the elvers ascend through the bristles. A cover is provided for the flume to protect the eels from predation and because they prefer to migrate in darkness.

Figure 3.31 shows details of a modern fishway for eels at the Robert H. Saunders Dam on the St. Lawrence River in Canada. This fishway surmounts a dam 28 m in height and was constructed first of wood in 1974. Natural brushwood was used in the flume to help the eels to ascend. The natural brushwood was later replaced by synthetic branches as shown in the diagram, for the reasons mentioned above, and the flume was replaced by a more durable double aluminum chute in 1981. The dimensions of each of the two flumes are 30 cm (12 in.) wide by 25 cm (10 in.) deep. The synthetic branches are placed in the flume, and the chain link, welded to the sides at intervals, is used for securing the branches as well as the bars or cleats welded to the bottom of the flume. The cleats also serve to reduce the velocity of the water. Here again, the depth of water is maintained at about 5 cm, which is sufficient to keep the branches thoroughly wet. The flume is placed in an unused ice chute so that it receives natural shade, and ascends the face of the ice chute in a zigzag pattern. Each leg of the chute rises at a slope of 12°, and there is a level section for the reverse bend at the end of each leg. Water is fed to the flume at the top from a pump, which also supplies the attraction water at the entrance by pipe.

While a great variety of widths and depths of flume have been used over the years, in general it is considered that a single flume 30 cm wide and 25 cm deep is capable of passing a run of 500,000 elvers annually, while a double flume such as that used at the Saunders Dam is capable of passing over 1 million eels annually without overcrowding. The fishway at the R.H. Saunders Dam has in fact passed more than 1 million eels in several of the years since it was built.

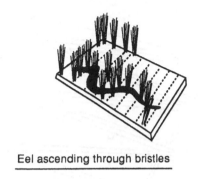

Eel ascending through bristles

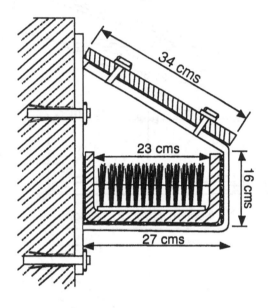

X-Section Through Fishway

Figure 3.30 Details of the eel fishway on the Staustufe Zeltingen/Mosel in Germany. (From Jens et al., *Arb. Dtsch. Fisch. Verbandes*, 32:1981. With permission of author.)

Pipes have also been used to pass eels, and with the plastic piping and synthetic bristles now available, this has proven to be a successful method of passing eels over low dams. De Groot and van Haasteren (1977) report on an installation of this type on the River Maas in The Netherlands. The pipe was 25 cm in diameter, placed vertically to a height of 5 m. It was filled with a synthetic brush (like a bottle brush) of the same diameter. From 15 to 30 l of water per minute were passed through the pipe, and the installation passed 1220 kg of elvers in 1976. At an estimated 700 elvers per kilogram, this amounted to approximately 850,000 elvers.

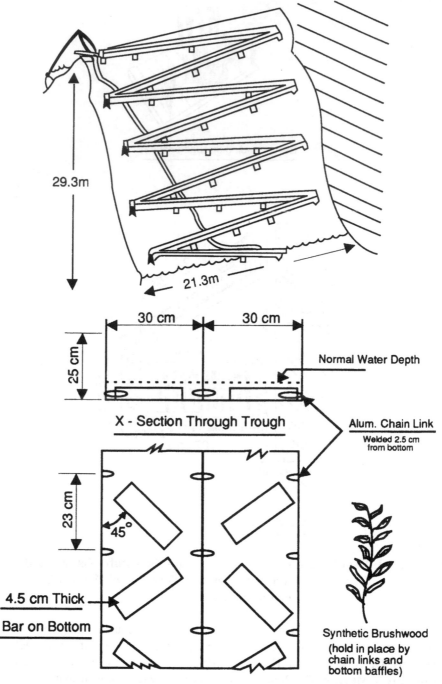

Figure 3.31   The Ontario Hydro R.H. Saunders Dam Fishway for eels on the St. Lawrence River in Canada. The perspective view (top) shows how it is placed in the unused ice sluice. (Drawn by M.C. Belzile.) Below, details of the double aluminum trough are shown along with the synthetic brushwood that lies in the bottom of the troughs for eel passage. The fishway is owned jointly by the Ontario Hydro and Ontario Ministry of Natural Resources.

This method was later adopted in New Zealand for use at a high dam, but there were unforeseen problems. The eels in New Zealand are of two different species, but their behavior is similar to those in Europe, and there seemed little risk in using the method developed in Holland. Mitchell (1984) describes the installation at the Patea Dam, which surmounts a height of 68 m, as shown in Figure 3.32. The selection of the pipe pass was made after considering the various alternatives and their associated costs. The pipe was selected as being the most economical. Several types of pipe were tested in the laboratory including corrugated iron, corrugated PVC, and smooth PVC. The smooth PVC, diameter 20 cm, was chosen on the basis of preference by elvers in the laboratory tests.

For the installation in the dam the pipe was filled with three bottle-type brushes in 1-m lengths with joints staggered. A small-diameter polyethylene tube was used to join the sections of the brushes. The total length of the pass is 250 m, and it lies partly next to the penstocks and partly over the exposed surface of the dam. The entrance to the pass is located near an outflow of "compensation" water, which runs at night when no flow is coming from the turbines. Since the eels migrate at night, this was an ideal combination for the fishway entrance. Provision was made for opening the pipe for inspection, and the flow down the pipe from the forebay was capable of regulation to get the best results.

An inspection made by Mitchell (1985) at the end of the migration for that year revealed some mortality in the pipe, although the pass was functioning earlier and passing elvers at a rate of 150/h. From a complete inspection it was found that many eels had died while ascending the pipe. It was surmised that the mortality occurred because of a combination of high temperatures (about 30°C) and lack of oxygen in the water. The lack of oxygen was caused by an increase in metabolic rate of the eels resulting from the higher demand caused by the high temperature. The elvers travel mainly at night, and it was thought that ascending this high dam took them two nights and a day. The suggested remedies were to shade the pipeline as much as possible, increase the water supply to the pipe, and possibly introduce fresh water at intervals throughout its length.

The reason given for using a pipe was maximum economy, but in comparison to an aluminum flume the cost saving must have been minimal, and many of the problems would have been lessened or more easily corrected with a flume. A direct cost comparison is difficult, since the full cost of the Patea Dam installation has not been published. However, the full cost of the Saunders Dam fishway was about $256,000 Canadian, and the aluminum flume was estimated at $58,000, over one fifth of the total. While the PVC pipe at Patea Dam probably cost less than the aluminum flume, the other costs, which include the pump, header tank, and entrance facilities, must be comparable. At the same time it should be noted that the flume at the Saunders Dam was only 140 m long; whereas, the pipe at Patea Dam was 240 m long and surmounted over twice the height.

Entrance conditions for an eel fishway are similar but not as critical as for a fishway for salmon and trout. The entrance must be placed near the migration point farthest upstream, as for salmon and trout. And for eels, some turbulence should be created near the point of entry to attract the eels. This can consist of a separate flow of attraction water emptying into the water near or at the point of entry, or sprayed over the entrance, or a combination of both.

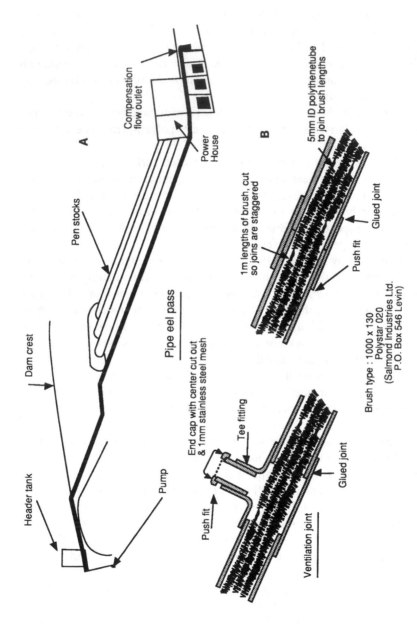

Figure 3.32 (A) Eel fishway in the Patea Dam in New Zealand. The dam is 68 m high, and the eel pipe is 250 m long. (B) Cut-away sections of the pipe show the three synthetic brushes in it, and the pipe "Tee" installed at intervals for ventilation.

The exit must be carefully located as far as possible upstream to prevent the young eels from being swept back over the dam or through the turbines.

3.18 FISHWAYS FOR OTHER CATADROMOUS SPECIES

As mentioned earlier, these species are a problem mainly in the southern hemisphere, although Pavlov (1989) deals with some species that migrate upstream as young in tributaries of the Caspian Sea.

In the southern hemisphere there is a concern for them in South Africa and Australia. In South Africa the species is freshwater mullet, and in Australia there are a number including golden perch, Australian bass, and barramundi. All of these enter fresh water and migrate upstream as juveniles, some as small as 2.5 cm in length.

Bok (1988) describes a new program in South Africa aimed at finding a suitable fishway for the freshwater mullet. He states that they have experimented in the laboratory with a weir type fishway that shows some promise. He states further:

> The weirs had a slope of about 40 degrees between pools. In addition, the weirs were set at an angle to the side walls which resulted in the water depth over the weir varying from *ca* 1-15 cm. The mullet invariably chose to swim over in shallow water where, of course, the flow rate was reduced. It is anticipated that different size fish will choose their desired depth/flow combination when swimming over these weirs.

He adds that other catadromous organisms, such as elvers and freshwater decapods, could use the very shallow weir depths for migration. So far the fishway has not been tested in the field, and no publication is available describing it in more detail.

In Australia, there is also a program commencing in the New South Wales Division of Fisheries on fish passage research. Mallen-Cooper (1988) reports that they are testing various vertical slot baffles at low heads of 10 cm per baffle and up, to try to find a head suitable for the target fish. The key to finding this head is knowledge of the swimming ability of the fish, and it is hoped that research in this aspect will proceed along with the practical experimentation with different types of baffles and fishways. It seems clear that no magical solution will appear. Whatever fishways eventually are favored will depend on their matching the requirements of the fish. This will mean lower gradients through the fishway, and as a consequence, longer fishways. It may be possible to design narrower fishways (such as the elver flumes described previously) and thus reduce both the structural cost and the water requirement.

3.19 LITERATURE CITED

Andrew, F.J. and G.H. Geen, 1958. Sockeye and Pink Salmon Investigations at the Seton Creek Hydroelectric Installation, Int. Pac. Salmon Fish. Comm. Prog. Rep. 73 pp.

Bell, M.C., 1984. Fisheries Handbook of Engineering Requirements and Biological Criteria, U.S. Army Corps of Engineers, North Pac. Div., Portland, OR. 290 pp.

Bok, A.H., 1988. Personal communication.

Bonnyman, G.A., 1958. Fishery requirements. In *Hydro Electric Engineering,* Vol. 1, Blackie and Son, London, pp. 1126–1155.

Clay, C.H., 1960. The Okanagan River flood control project, 7th Tech. Mtg., Athens, Int. Un. Conserv. Nature, Brussels, Vol. IV, pp. 346–351.

Committee on Fish Passes, 1942. Report of the Committee on Fish Passes, British Institution of Civil Engineers, William Clowes and Sons, London. 59 pp.

Committee of the Hydraulics Division on Hydraulic Research, 1942. Hydraulic Models, Am. Soc. Civ. Eng., New York, 110 pp.

Conrad, V. and H. Jansen, 1983. Refinements in Design of Fishways for Small Watershed, Fish. Oceans, Canada, Scotia Region, 26 pp.

Conrad, V. and H. Jansen, 1987. Personal communication.

Cooper, A.C. and W.E. Boresky, 1953. Report on Model Studies of Proposed Fish Protective Facilities for Seton Creek Dam, Int. Pac. Salmon Fish. Comm. Manu. Rep. 77 pp.

De Groot, A.T. and L.M. van Haasteren, 1977. The migration of young eels through the so-called eelpipe, *Visserij, Voorlichtingsblad voor de Nedrlandse Visserij,* 30(7).

Dept. of Fisheries, Canada, and Int. Pac. Salmon Fisheries Comm., 1955. A Report on the Fish Facilities and Fisheries Problems Related to the Fraser and Thompson River Dam Site Investigations, Dept. Fish., Vancouver, B.C. 102 pp.

Elling, C.H. and H.L. Raymond, 1956. Fishway Capacity Experiment, 1956, U.S. Fish & Wildlife Serv. Spec. Sci. Rep. No. 299. 26 pp.

Fulton, L.A., H.A. Gangmark, and S.H. Bair, 1953. Trial of a Denil-Type Fish Ladder on Pacific Salmon. U.S. Fish & Wildlife Serv. Spec. Sci. Rep. Fish No. 99. 16 pp.

Jens, G. et al., 1981. Function, construction and management of fishpasses, *Arb. Dtsch. Fisch. Verbandes,* Heft 32.

Katopodis, C. and N.C. Rajaratnam, 1983. A Review and Laboratory Study of the Hydraulics of Denil Fishways, Can. Tech. Rep. Fish. Aquatic Sci. No. 1145. 181 pp.

Lander, R.H., 1959. The Problem of Fishway Capacity, U.S. Fish & Wildlife Serv., Spec. Sci. Rep. No. 301. 5 pp.

Larinier, M., 1983. Guide pour la conception des dispositifs de franchissement des barrages pour les poissons migrateur, *Bull. Fr. Piscic.,* July.

Larinier, M., 1988. Personal communication.

Larinier, M. and D. Trivellato, 1987. Hydraulic Model Studies for Bergerac Dam Fishway on the Dordogne River, La Houille Blanch No. 1/2.

Long, C.W., 1959. Passage of Salmonoids through a Darkened Fishway, U.S. Fish & Wildlife Serv. Spec. Sci. Rep. No. 300. 9 pp.

Mallen-Cooper, M., 1988. Personal communication.

McGrath, C.J., 1955. A Report on a Study Tour of Fishery Developments in Sweden. Fish. Br., Dept. Lands, Dublin. 27 pp.

McLeod, A.M. and P. Nemenyi, 1939–1940. An Investigation of Fishways, Univ. Iowa, Stu. Eng. Bull. No. 24. 66 pp.

Mitchell, C., 1984. Patea Dam Elver Bypass, Recommendations to the Consulting Engineers, Fish. Res. Div., Min. Agric. Fish., Rotorua, N.Z.; 1985. Report on the Elver Pass at Patea Dam, Informal Report, Min. Agric. Fish, Rotorua, N.Z.

Pavlov, D.S., 1989. Structures Assisting the Migrations of Non-Salmonid Fish: U.S.S.R., FAO Fisheries Tech. Pap. No. 308, Food and Agriculture Organization of the United Nations, Rome. 97 pp.

Pretious, E.S., L.R. Kersey, and G.P. Contractor, 1957. Fish Protection and Power Development on the Fraser River, Univ. B.C., Vancouver. 65 pp.

Tesch, F.W., 1977. *The Eel,* Chapman & Hall, London. 423 pp.

Thompson, C.S. and J.R. Gauley, 1964. U.S. Fish & Wildlife Serv. Fish Pass. Res. Prog., Prog. Rep. No. 111. 8 p.

Von Gunten, G.J., H.A. Smith, and B.M. MacLean, 1956. Fish passage facilities at McNary Dam. Proc. Pap. No. 895, *J. Power Div. A.S.C.E.,* 82, 27 pp.

Zeimer, G.L., 1962. Steeppass Fishway Development, Alaska Dept. Fish Game Inf. Leafl. No. 12. 27 pp.; March 1965 Add. 5 pp.

4   FISH LOCKS AND FISH ELEVATORS

4.1 DEFINITIONS AND HISTORY

Fish locks and fish elevators have been grouped together in this chapter not because of any resemblance in their method of operation, but because they both represent alternatives to conventional fishways for passing fish over dams. A fish lock is defined as a device to raise fish over dams by filling with water a chamber, which the fish have entered at tailwater level or from a short fishway, until the water surface in it reaches or comes sufficiently close to forebay level to permit the fish to swim into the forebay or reservoir above the dam. It is similar to a navigation lock, and indeed, fish have been known to pass upstream through navigation locks on many occasions. A fish elevator is defined as any mechanical means of transporting fish upstream over a dam, such as tanks on rails, tank trucks, buckets hung on a cable, etc., and will in this text include the means of collecting and loading the fish into the conveyance.

Both fish locks and fish elevators have a much shorter history than fishways. A Mr. Malloch, of Perth, Scotland, has been credited with evolving a scheme similar to modern fish locks around 1900, but evidently this was too early for the idea to be accepted, as no immediate applications followed. The period of the initial application of fish locks and elevators on a practical scale corresponds to a period when dams were being designed that were much higher than any previously conceived. This was in the 1920s. As long as dams were less than about 50 ft (15 m) in height, conventional fishways were considered to be unduly expensive. But as dams higher than this became more common, and some were envisaged on salmon streams to heights of 300 ft (100 m), there was more incentive to find alternative methods for providing adult fish passage. Another factor besides economy was instrumental in the development of these devices. This was the fear that fish would not be physically capable of ascending fishways over high dams. This belief is still prevalent today, and is probably justifiable on the basis that so little is known of the physical capabilities of the fish. More is being learned each year, however, and fishways have been constructed at dams of increasingly greater heights so that the earlier fears have to a certain extent been reduced.

The lack of data on the stresses imposed on fish by their ascent through fishways at dams has undoubtedly been instrumental in the development of both fish locks and fish elevators, because both of these devices require less effort on the part of the fish than conventional fishways. This does not say that they are recommended, however. Because there is also a lack of data on the stresses imposed on the fish by passage through these devices, the advantage of less effort is more apparent than real.

Nemenyi's bibliography (1941) refers to articles describing an experimental fish elevator tested on the White Salmon River in Washington about 1924, and to a patented fish lift (or lock) tried on the Umpgua River in Oregon in 1926. The earliest similar installation in Europe is noted in the same bibliography as an elevator at Aborrfors, Finland, first described in 1933. This was followed, apparently within a few years, by several other installations in Finland, and by an elevator on the Rhine River at Kembs.

About this time, interest in North America turned to an installation at Baker Dam in the State of Washington, which utilized a cable and bucket system to pass fish over a total height of almost 300 ft. This contrivance was hailed at the time as the answer to the problem of passing fish over all high dams, but it is interesting to note that it has recently been replaced by a new trapping and trucking operation. The decline of the salmon runs to the Baker River in the intervening period of about 30 years cannot be attributed wholly to the effects of the original hoist, but to the high mortality of downstream migrant juvenile fish at the dam, which was measured and recorded by Hamilton and Andrew (1954). However, the replacement of the system for adult fish passage at considerable cost is assumed to be evidence of dissatisfaction with the original method. Bonneville Dam, which was also constructed in the 1930s, was furnished with large fish locks in addition to the elaborate conventional fishways described previously. These fish locks were mainly experimental, but have served a useful purpose as well, as will be described later.

There are no publications to indicate that there was further consideration of fish locks or elevators on any extensive scale until the period following the end of World War II, 15 to 20 years later. This does not mean that there was a complete cessation of interest in these devices in the interim, however. In the period 1939 to 1943, following construction of Grand Coulee Dam in Washington State, a temporary trapping and trucking operation at Rock Island Dam, on the Columbia River some distance downstream, was successfully used to transport several thousand adult salmon from the latter dam to new spawning grounds. Fish and Hanavan (1948) report the details of this operation, which was necessitated by the loss of the original spawning grounds above Grand Coulee.

In the same period, fish elevators were developed as a practical facility for fish passage at high dams in the U.S. and Canada. The White River trucking operation to provide fish passage over Mud Mountain Dam in Washington State was one of the first of these. Trapping and trucking operations were used on the Sacramento River in California in the years following 1943. Moffett (1949) describes the results of these operations at Keswick and Balls Ferry below Shasta Dam.

Trapping and trucking installations were placed at Mactaquac Dam on the Saint John River on the east coast of Canada in 1967, and at the Essex Dam on the Merrimack River on the east coast of the U.S. in 1981. The latter was intended mainly to provide passage for shad and alewives, and was based on the success of an earlier installation on the Holyoake River at Holyoake, Mass.

After World War II, development of fish locks as practical fish-passage facilities accelerated in Europe, starting with the Leixlip development on the River Liffey near Dublin, Ireland, in 1949 to 1950, and continuing in Scotland, Ireland, and Russia up to the present time. In Scotland and Ireland, development centered round the Borland lock which will be described later. In East Europe, the main country to develop fish locks and fish lifts has been Russia. Brief references to the early work were made by Klykov (1958) and Kipper (1959). These have been followed by many publications in Russian, which have been summarized by Pavlov (1989), who describes in detail the installations on the rivers of the basins of the Caspian, Azov, and Black Seas. These rivers include the Volga, Don, and Kuban, on which extensive hydroelectric and other water-use projects have been constructed since World War II.

In the following sections we discuss some of these installations in more detail.

4.2 FISH LOCKS

The first of the modern fish locks in Europe was constructed in 1949 at Leixlip Dam on the River Liffey near Dublin, Ireland. It was based on a design by J.H.T. Borland, of the firm of Glenfield and Kennedy, and subsequent installations have been called Borland fish locks or fish passes. Since that time, more than a dozen have been built in Scotland and Ireland, surmounting dams of up to 200 ft (61 m) in height. They have passed salmon numbering up to 8000 annually, and are accepted as operating satisfactorily in maintaining the runs of salmon and trout.

In North America, a Borland lock was constructed in Ontario on the Haines River at Thornbury. It successfully passes similar-sized runs of Great Lakes chinook salmon and rainbow trout over its height of 7.3 m (24 ft).

According to Quiros (1988), Borland locks have been installed at a dam in South America, the Salto Grande on the Uruguay River. At 30 m (100 ft) in height, it is among the highest installations in the world. It is, however, less satisfactory for the species and numbers of fish to be passed, as will be described later.

While operating over a considerable range of heads, Borland locks are generally used at dams of more than 10 m (30 ft) in height. At dams lower than this, a fish pass of standard design is considered to be more effective and economical.

Most Borland locks are similar in design to the one shown in Figure 4.1, and a brief discussion of its operation as described in *British Engineering* (Anonymous, 1950), by the Glenfield and Kennedy Firm, will illustrate generally the operation of most of them. With the lock controls adjusted to operate on a 1-h cycle, starting with the fish entrance sluice-gate open, the adjustable weir in the top chamber is set to permit about 10 cfs to flow in and down the sloping chamber. This flows out the fish entrance sluice-gate and attracts the fish into the bottom chamber. At the end of 25 min the fish entrance sluice-gate closes, and the flow of 10 cfs is now collected in the bottom chamber for 5 min. At this time the fish exit sluice-gate opens, increasing the rate of flow and causing the bottom chamber, the sloping chamber, and the top chamber to fill in about 5 min. The fish rise to the top chamber as the water level rises, and are free to enter the reservoir for 25 min after the top chamber is filled. They are encouraged to leave by the flow inward through the exit gate induced by opening the bypass valve beside the bottom chamber. At the end of 25 min the fish

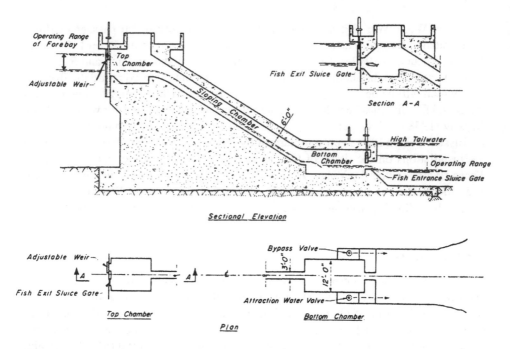

Figure 4.1   Schematic plan of a typical Borland fish lock. Operation of the lock is described in the text.

exit sluice closes, and 5 min later the fish entrance sluice in the bottom chamber opens and the cycle is repeated. The last 5-min period before the entrance gate is opened allows time for most of the sloping chamber to empty through the bypass valve, so that when the entrance gate opens, the water level is low enough to avoid excessive turbulence as the bottom chamber empties. The attraction water valve is used to increase flow into the tailrace adjacent to the fish entrance. This water is taken directly from the forebay and is used to increase the attraction to the fish entrance when needed as a result of tailrace conditions.

Borland locks are generally constructed to operate automatically on various cycles such as 1 h, 3 h, etc. On inspection of several of those in Scotland, however, the author found that manual operation was being used, particularly during periods when runs were not at their peak.

Borland locks are also used in many cases to pass salmon, smolts, and kelts (spent spawners) downstream. In some cases they provide the only means of egress from the reservoir when the turbine intakes are screened and no spill is taking place. The Salmon Research Committee of the Scottish Home Department and North of Scotland Hydroelectric Board reported in 1957 (unpublished) that as many as 458 kelts descended Torr Achilty fish lock, and in one season 13,943 smolts were recorded descending the Meig Dam fish lock.

In addition to the Borland type, several other types of fish locks have been installed in Europe. One such has been installed on the River Shannon at Ardnacrusha Dam in Ireland.

The Ardnacrusha fish lock differs from the typical Borland lock described mainly in that the sloping chamber is replaced by a vertical cylindrical chamber, as

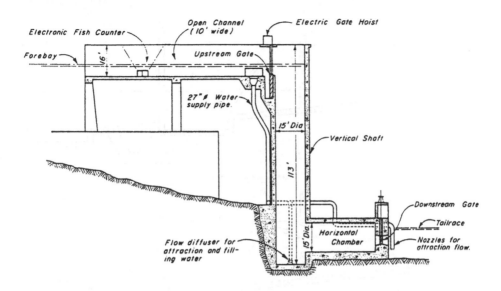

Figure 4.2   Sectional elevation of the fish lock in the Ardnacrusha Dam on the River Shannon in Ireland. This lock has an approximate average working head of 94 ft.

shown in Figure 4.2. Fish enter the base of this cylinder or shaft, which is 15 ft in diameter, and are raised to forebay level as the shaft fills after the downstream gate is shut. They then enter the headrace canal by means of a horizontal open channel spanning the distance from the top of the cylindrical shaft to the top of the dam. The flow, which attracts the fish into the entrance to the lock at the base of the dam, is supplied by a 27-in.-diameter pipe leading from the horizontal open channel near the top of the main shaft This pipe branches as shown in Figure 4.2, with one branch supplying water at the base of the shaft through a disperser and the other discharging through nozzles outside the entrance gate to attract fish. There does not appear to be any bypass valve similar to that in the standard Borland fish lock previously described, but it is assumed that when the fish rise to the surface of the water in the main shaft, they are encouraged to swim out of it into the headrace by the velocity induced in the horizontal channel by the discharge through the 27-in.-diameter pipe.

The cycle of operation for the Ardnacrusha fish lock is stated as 4 h, with 2 h for collecting fish in the horizontal chamber and approximately 70 min at the full stage, when the fish are passing from the lock into the headrace. This cycle is shortened to 2 h during the peak of the eel migration, which is extremely heavy.

The installation in Orrin Dam (Scotland) is also unique, in that it consists of a battery of four Borland locks, rather than a single one. Orrin Dam is approximately 200 ft high and forms a storage reservoir which is operated with a fluctuation in surface level of 70 ft. The top chambers of the four locks are distributed at different elevations over this range, so that a fish lock is always in operation and fish can pass upstream regardless of reservoir level. No doubt this has added considerably to the cost of transporting fish over this dam and has thereby reduced considerably any economic advantage this method of fish passage might have had over alternate methods. The Orrin fish locks are also unique in that the sloping chamber has been

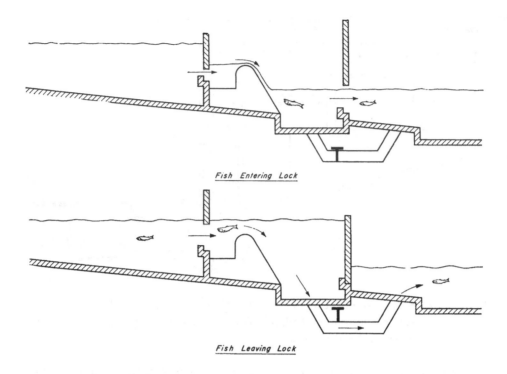

Fish Entering Lock

Fish Leaving Lock

Figure 4.3 Two views showing schematically the operation of a fish lock used in Holland. (After Deelder, 1958.)

extended below the level of the bottom chamber to form a cushioning pool in which the energy of the water passing down the sloping chamber is dissipated in turbulence.

On the European continent there is a small lock on the River Meuse in Holland described by Deelder (1958). It surmounts a weir at Lith having a height of approximately 4 m and is shown schematically in Figure 4.3. It appears to be similar in principle to the Borland lock except that the bottom chamber and sloping chamber are open at the top. As a result, the water rises in them to the full height of the reservoir above the weir after the fish entrance gate is closed. This obviously is practical only at a low-head installation. For higher heads, a closed lower chamber would probably be required for structural and economic reasons. Data on the dimensions, flows, and operating cycle are not known, but it is stated that all fish observed through a special window incorporated in the structure ascended with ease, including the smallest stickleback.

In Russia, development seems to have taken the form of a movement into fish locks, as defined in this text, in the 1950s and early 1960s, followed by a gradual change, first to fish elevators at dams, then to fish traps and barging or trucking facilities located downstream of the dams. The fish locks were located at the Tzimlyanskij project on the Don River (built in 1955), the Volzhskaya power plant on the Volga River (1961), and the Volkhovskij project on the Volga River (1967).

The Tzimlyanskij fish lock is shown in Figure 4.4. Because it is intended to pass species other than salmon, it has several features not included in the Borland locks. Its auxiliary water attraction system, fish crowder for inducing fish to enter the lock chamber, and lifting basket

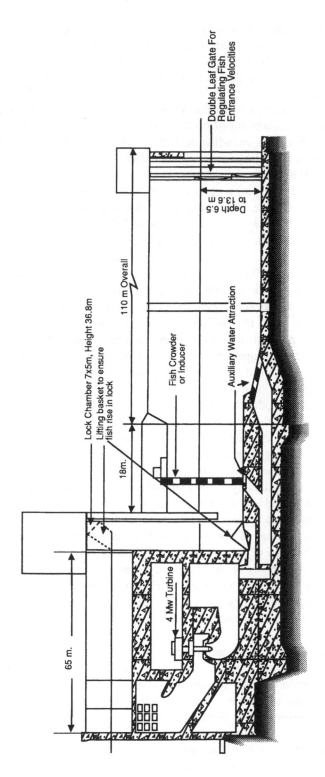

Figure 4.4 The Tzymlyanskij Fish Lock on the Don River, former U.S.S.R., first built in 1955 and reconstructed in 1972. (After Pavlov, D.S., 1989.)

to ensure they ascend the lock are all extras. This fish lock has an operating cycle of 2.5 to 3.0 h. Dimensions of the lock vary, but in general it is about 5 to 6 m in width with a minimum water depth in the fish intake channel of 6.5 m and in the outlet channel to the reservoir of 2 m. The system was improved in 1972, and the present fish lock passes sturgeon, bream, carp, wels, sheatfish, and other species.

The Volzhskaya installation on the Volga River is a similar but improved version. It has passed a maximum of 60,000 specimens of sturgeon (in 1967) and large numbers of other species including the Caspian inconnu and herring. It has two locks side by side, one collecting fish while the other is passing them through, and the locking cycle is stated to be 1.5 to 2.0 h.

As noted, Borland type locks have been installed in South America on the Uruguay River at Salto Grande. The dam is some 30 m in height, and spans the river between Argentina and Uruguay. It was completed in 1982, and to date the operation has met with a number of difficulties which have been only partly rectified. The dam contains two locks, one at each end of the spillway, as shown in Figure 4.5. Mean annual flow of the river is 4500 m³/s, while flows in the lock are only between 0.5 and 1.0 m³/s. The only entrances to the locks are on the spillway side, and as one would expect, the fish experience a great deal of difficulty in finding them.

The main species using the facility are the characins, *Salminus maxillosus* (dorado), *Prochilodus platensis* (curimbata), and *Leporinus obtusidens* (leporinus). Their behavior in migrating upstream was not clearly understood, and adequate provision was not made in the dam for collecting and passing them. It would seem, from Quiros' description of their behavior after the dam was built, that they gather below the powerhouse on both sides of the river as well as below the spillway gates. A collection gallery across the face of the powerhouse leading to standard pool and weir fishways on each bank would seem to offer a partial solution to the problem. Operation of the spillway gates in a manner that benefits fish migration would also help. In view of the imminent installation of other Borland locks and impending construction of new dams on many of the rivers of the region, the problems encountered at Salto Grande would appear to be worthy of careful study.

The strongest argument usually given in favor of fish locks is their economy, both in capital cost and in cost of operation. No direct comparison of costs is given in the literature, but because of the smaller size of the lock structure, compared to a conventional pool type fishway of equal capacity, there seems little doubt that the capital cost of the lock would be less, particularly for higher dams.

Another advantage of fish locks, which is often cited, is the economy in use of water. However, the typical Borland lock appears to have an average discharge of 25 cfs; this is sufficient to operate a conventional pool type fishway 6 to 8 ft wide, which probably has a larger fish-handling capacity than a fish lock. It is logical to assume that the attraction flow out of the fish lock entrance must be equivalent to the flow out of a conventional fishway entrance to produce an equivalent level of attractiveness, so that no advantage in water economy appears to be possible during the part of the cycle when fish are entering. During the part of the cycle when fish are leaving the lock, almost the same logic applies. The greater the flow entering the upper chamber, the more success in inducing fish to clear the chamber (within limits), so that a flow equivalent to the minimum conventional fishway is probably required. These two portions of the cycle take up a large proportion of the total cycle time, so

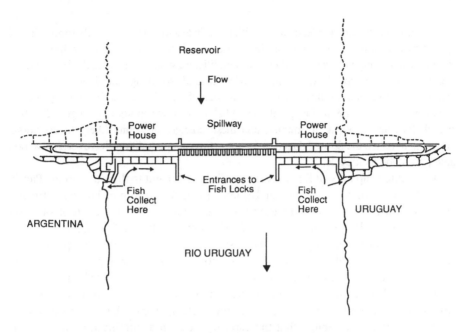

Figure 4.5   The Salto Grande Dam on the Uruguay River with Borland fish locks at each end of the spillway. The entrances to the fish locks are on the spillway side where shown. Exit from the locks is in the reservoir close to the spillway gates and turbine intakes.

that any economies in water flow effected in the balance of the cycle are almost insignificant.

It is suggested, therefore, that the commonly cited overall advantages of the fish lock should be regarded with caution, particularly if the efficiency has not been fully assessed. There does not appear to have been any attempt so far to assess efficiency by determining if any delay or loss of adult migrants occurs at any existing installations in Europe. There has been a report, on the other hand, of losses by injury in the Borland locks in Scotland that appeared to have been substantiated by direct observation. The South American installation, however, demonstrates fully the limitations of this type of lock, when applied to other species and large rivers with large numbers of migrants.

The most apparent weakness in the fish locks tested on the Columbia River has been the inability to clear all fish from the upper part of the lock chamber in the limited time permitted by the preset cycle. In the case of locks in Scotland, where this same problem has occurred, fish remaining in the lock after the exit gate has closed, have occasionally been washed down the sloping chamber and injured in the lower chamber either by abrasion, or by impact on the chamber walls or on the entrance gate. This led in part to the continuation of the sloping chamber below the bottom chamber at Orrin Dam, to provide a cushioning pool at the base. While the possibility of direct injury to some of the fish is a serious matter, the fact that others might have to repeat passage through the lock cycle several times before passing upstream is considered to be equally serious. The capacity of the fish lock is definitely limited by the volume of the lower chamber or collection pool, and the length of the time cycle. It can be even further limited by the necessity to pass "repeaters."

It is believed that any tendency for the fish to hesitate leaving the upper chamber of a lock is probably due to the same condition that causes a reduction in the rate of migration in a conventional pool type of fishway when flow conditions are changed suddenly. For example, when flow in a weir type fishway is increased sufficiently to change the flow over the weirs from plunging to streaming flow, migration has been observed to lessen in intensity or cease altogether for a short time. In the operation of a fish lock, flow conditions change radically as the lock fills, and again as it empties, and these changes are quite likely to discourage movement of fish in the desired direction for some time after conditions have reached a more stable part of the lock cycle.

The limited capacity of fish locks is a definite deterrent to their use on the Pacific Coast of North America. A typical Borland lock, for example, has an estimated volume of water in its lower chamber at maximum tailwater elevation of about 2400 ft^3. This would accommodate 600 fish, allowing 4 ft^3 per fish. On a 3-h cycle, the bottom chamber capacity limits the pass to a maximum of 200 fish per hour, or considerably less than the capacity of a conventional fishway with pools 6 ft wide by 8 ft long by 6 ft deep. An increase in the lock size or addition of more locks increases the cost of fish facilities to a point where it is higher than that of a conventional fishway. The best application for fish locks therefore appears to be at dams where the runs of fish are small enough that the capacity of even a minimum-width conventional fishway is not required. Advantage can then be taken of the saving in cost of the lock over the conventional fishway.

The description of the Borland lock would not be complete without a word as to its location in a dam, in case the reader wishes to consider its application. Figure 4.6 shows a typical layout of a Borland fish lock on a small river to pass a run of salmon over a hydroelectric dam. The same rules apply to location of lock entrance as for a conventional fishway entrance. The designers of the Scottish and Irish installations have not resorted to elaborate auxiliary water systems or to collection galleries at the entrance, however. They have depended on guiding fish to the entrance by use of a rack or screen angled across the tailrace. The simple auxiliary water system described earlier in this section provides attraction for the fish by adding a small flow adjacent to the entrance gate. The structural economy as compared to a conventional pool fishway will be noted. It will also be noted that it is comparatively easy to locate the gate to the top chamber where it has a reasonable chance of attracting downstream migrants.

It will be noted, however, that the Borland lock has been used successfully in this way on comparatively small rivers with small runs of salmon. Their design can not be extrapolated to large rivers with large stocks of migratory fish without detailed consideration of the problems of attraction, entry, and capacity, as outlined in Chapter 3.

4.3 FISH ELEVATORS

Fish elevators have been used in preference to fish locks in North America to solve the problem of passing small runs of salmon and larger runs of shad and alewives over high dams. These elevators have mainly taken the form of trapping and trucking operations in recent years. In Russia they have been used in place of fish locks since the 1960s, to take care of the multitude of other species including the sturgeon on the Volga, the Don, and the Kuban Rivers.

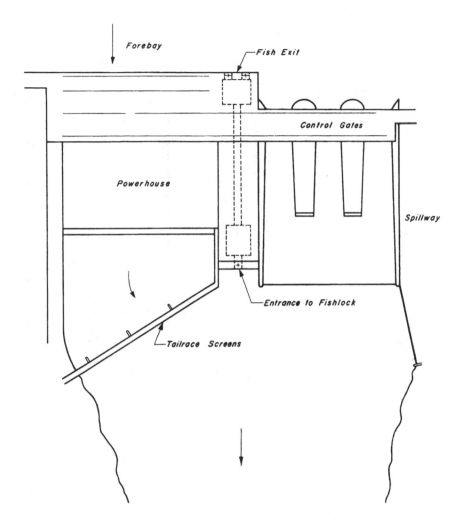

Figure 4.6 Typical layout of a hydroelectric dam with a Borland fish lock.

Some of these facilities involve trucking the fish, and others barging them or dumping them in the forebay of the dam.

The newer North American installations are on the East Coast at Mactaquac on the Saint John River in Canada and at Essex Dam on the Merrimack River in the U.S. Both of these installations have large numbers of shad and alewives as well as smaller numbers of Atlantic salmon. An installation that is typical of facilities built in North America is shown in Figure 4.7. After passing up a standard fishway, the fish enter a holding pool. The final fishway weir at the entrance to this pool is equipped with a finger trap (Figure 4.8) to prevent fish from returning downstream. This pool is supplied with water through a diffusion grating in the floor. The grating and diffusion chamber underneath it are designed on the same principles as previously described for fishways at dams. This flow combines with flow from the brailing pool adjacent to it to make up the required fish pass flow.

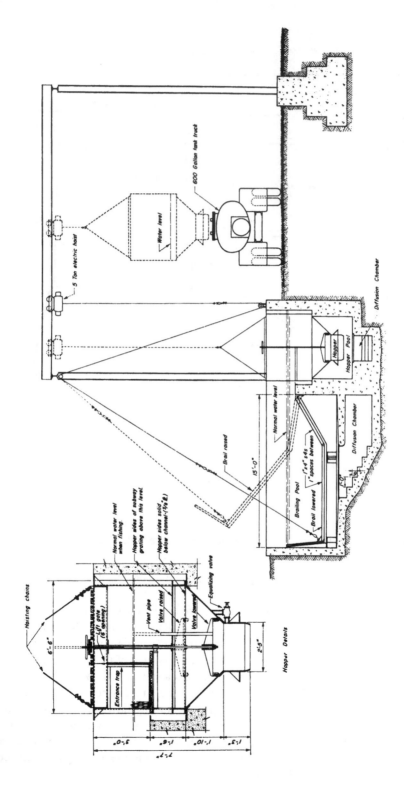

Figure 4.7 Sectional elevation of a typical fish elevator installed on the Pacific Coast of North America.

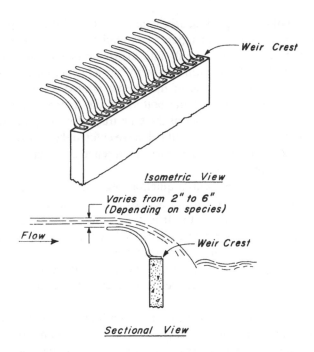

Figure 4.8 **A finger trap of the type commonly used in North America to prevent return of adult fish downstream after ascending a weir into a holding pool.**

The brailing pool lies immediately adjacent to the holding pool. It is also supplied with water through a diffusion grating in the floor. This water flows out of the brail pool into the holding pool through a connecting gate, thus encouraging the fish to enter the brail pool when the gate is open. This opening is V-shaped in plan, which discourages the fish from leaving the brail pool after entering. The brail pool has a false bottom, constructed of wood slats, which pivots on pins set along the edge nearest the gate leading to the hopper pool. The false bottom can be raised to a tilted position by a winch, cable, and bridle attached to the side opposite to the pins as shown in Figure 4.7. As it swings up in an arc, the fish are forced into the entrance to the hopper pool. The hopper is in place in the hopper pool when the fish enter, and water is flowing up through a third diffusion grating in the floor of this pool This water then flows out the entrance gate to the hopper, and gives added attraction for the fish to enter the hopper. Two vertical combs, composed of round bars spaced to fit between the wood slats of the false floor of the brail, prevent the fish from lodging in the corners of the brail pool as the false floor is tilted up.

When all fish have been cleared from the brail pool into the hopper, the gate in the hopper is closed, the valve in the bottom is closed, and the hopper is raised by an electrically driven winch. The water level in the hopper then drops to the top of the tapered steel-plate base portion shown in Figure 4.7, and the fish are confined in this limited space. The hopper is, however, immediately raised and moved along the overhead track to a position over the tank truck, where it is lowered into place on the truck.

The upper part of the hopper is square in plan, so that it fits the hopper pool when lowered into it. This square section tapers to a round opening at the base. The lower part of this tapered transition fits snugly into the round opening or hatch in the top of the tank truck. The tank truck is full of water when a hopper load of fish is lowered into it. The small valve on the side of the hopper is next opened to equalize the pressure on the main valve closing the bottom of the hopper. This valve is then opened, after which water is bled out of the tank truck, lowering the water level in the hopper until it is empty and water and fish have entered the tank on the truck. The hopper is then removed, the tank truck cover is fastened down and the truckload of fish is ready for transport upstream.

The tank truck is equipped with a quick-acting release gate in the rear, which permits the fish and water to be sluiced out at the releasing site very rapidly. This reduces the possibility of injury to the fish in the unloading operation. Figure 4.9 shows this gate. The release mechanism is tripped by means of the lever on the left side, which disengages the cam rollers from the cams at each side of the gate, allowing the two heavy-duty springs to pull the gate fully open very quickly. The gate, which is sealed with a rubber gasket, has a clear opening a little less than 2 ft square.

The best method of releasing the fish has been found to be by depositing them directly from the tank truck into the receiving water. If this is not possible because of the inability to move the truck close enough to the water's edge, or if the receiving water is so shallow that injury might be sustained by the fish, then a steep wooden or metal chute can be used from the truck to the water. It is preferable that the chute be set at such a slope that the fish pass down it surrounded by water from the tank, rather than having them slide down the bare chute.

There is a trend toward automating these facilities to save on manpower and thus on operating costs.

The Russian installations differ from the North American ones because of the widely different mix of species they have to handle. Figure 4.10 shows a section through a typical installation built in 1969 on the Volga River at the Saratovskij project. It passes sturgeon, herring, carp, bream, and other species over the low dam and it is located between the power station and the dam. It is noteworthy because of its large size. The fish entrance is 172 m (564 ft) long and 8 m (26.24 ft) wide. It is assumed the entrance is in the form of a collection gallery. In operation, when sufficient fish enter the chute, the inducer or crowder is lowered and starts to move toward the dam. As it passes the grating over the lower draft tube, the gate to this tube is raised to the upper position, shutting off direct flow to the hopper chamber and directing it to the floor grating. The separating screen is then lowered, and the fish are raised in the hopper and moved over a large screened tank at forebay level, where they are released upstream.

Pavlov (1989) states that the turbine is of 9.5 Mw capacity, but he does not state whether all or part of the flow is utilized for the fish facility, or whether this turbine is specially installed for the facility or just one of the normal turbines at the dam.

Figure 4.11 shows a typical floating fish trap, which the Russians have used recently as part of a system of trapping and transporting fish over dams on the Keban, Don, and Volga Rivers. This particular fish trap is located at Kochetovskij on the Don River. It consists of a floating, non-self-propelled barge, which is anchored in place

Figure 4.9 Quick-acting gate on rear of tank truck for release of adult fish above dam.

by four piles as shown. It has a fish entrance channel 63.9 m (209.5 ft) long and 8 m (25.4 ft) wide. Overall, it is 13 m wide, which allows it to be transported through the navigation locks in the dams if needed at another location. It is supplied by nine axial-flow pumps on the end and sides to provide attraction water where needed.

In operation, the period of fish attraction and accumulation lasts from 1.5 to 2.0 h. Then the crowder or inducer is put down at the entrance and part of the attraction water is shut off, so as to produce a velocity of 0.4 to 0.5 m/s (1.3 to 1.6 ft/sec) in the chute. The crowder is then moved toward the container, concentrating the fish over a screened lifting device, which then lifts them to the transportation chute of the container vessel. The container vessel is self-propelled, with a cabin in the center containing control panels for navigating and locking and unlocking itself to the main fish trap. This vessel transports the fish upstream and releases them in the reservoir. The same type is used in conjunction with a trucking operation to move fish over the Rizhskaya Dam on the Daugava River. In that installation, an electrical guiding system is used to direct the fish to the fish entrance of the trap.

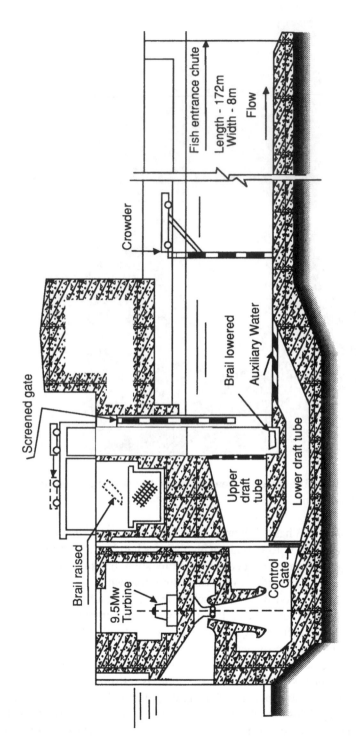

Figure 4.10 The Saratovskij fish lift or elevator on the Volga River, Russia, built in 1969. (After Pavlov, D.S., 1989.)

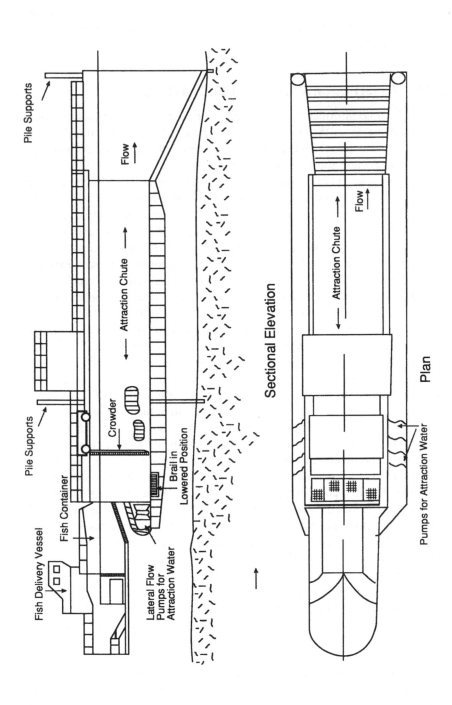

Figure 4.11 The Kochetovskij floating trap and barge for transporting fish on the Don River. (After Pavlov, D.S., 1989.)

The fish traps have the advantage of being mobile and flexible, with the ability to attract very small fish on local migrations as well as more mature sturgeon and other species. At some installations, they claim to have passed more than a million fish of all sizes annually.

4.4 COSTS OF FISH LOCKS AND FISH ELEVATORS

There has been no information published on the costs of the fish lock installations in modern dams. Since they are often constructed within the dam, it is difficult to deduce the cost from the available sketches and plans without knowing how much of the cost of the structure should be chargeable to fish facilities and how much to the dam itself. The upper and lower chambers of the lock, the steel liner of the sloping chamber, and all lock hydraulic and electrical controls, however, are definitely parts of the fish facilities and chargeable as such. In the case where a vertical shaft outside of the dam instead of a sloping chamber within the dam is used, such as at Ardnacrusha, the entire cost of this shaft should also be chargeable to the fish facilities.

Keeping all this in mind, and making only a general comparison with existing structures, for dams less than 60 ft (20 m) high, it would appear that fish locks would be about equal in cost to a fish elevator of the trapping and trucking type, excluding the cost of a barrier dam. In other words, the cost of installing a fish lock in a low dam would about equal the cost of trapping and trucking facilities installed at the base of the dam with a comparable fish entrance. For dams about 60 to 200 ft (20 to 61 m) high, it is probable that the cost of a fish lock will increase sufficiently to equal the cost of a trapping and trucking operation including a modest barrier dam or rack under favorable foundation conditions. This would be particularly true if an excessive drawdown of the forebay level necessitated use of more than one lock, as is the case at Orrin Dam in Scotland. In this range, therefore, costs of fish locks and fish elevators are fairly comparable, but the elevator is likely to be more efficient because of the presence of a barrier dam to guide the fish to the traps. For dams over 200 ft (61 m) high, it is likely that trapping and trucking facilities become much more favorable as regards initial cost, although operating costs are always higher, as will be pointed out later. This very general comparison, based on direct observation of both types of facilities in operation, will, it is hoped, give some impression of the scale of costs involved with fish locks, and will perhaps be more clear after reference to the more definite statements on costs of trucking operation that follow.

Starting with the tank trucks, the tank itself, with all piping, valves, etc., and two circulating pumps, can be supplied at the present time for costs varying from about $17,000 for a 300-gal tank to about $28,000 for a 1000-gal tank. The truck chassis cost will vary through a greater range, depending on tank capacity, but a good estimate would be from $20,000 to 50,000 for the range from 300 to 1000 gal. (These prices are all in 1987 U.S. dollars.)

While the cost of operation is a consideration in any type of engineering structure, it is particularly important in the case of a trapping and trucking installation. At least one employee is required on a full-time basis during the migration period to drive each truck, and two could be required for each truck if they are operated on two shifts per day. An additional man could be required for the loading operation, depending on the degree of automation installed. Assuming that the migration occurs over a six-month period, and only one truck is used, a minimum of one half man-year is required.

Additional part-time help might be required for trapping and loading, and shift work during the peak of the migration could increase the total labor costs to about $25,000. Truck operating and replacement costs could total $5,000 annually per truck. Repairs and maintenance to the structures will vary with the site, but could amount to a further $25,000 annually on the average. Annual charges for operation and maintenance could therefore total $60,000 for a typical minimum-sized trucking operation, or 5% of the capital cost, exclusive of any allowances for depreciation and interest on the investment.

4.5 COMPARISON OF FISH LOCKS AND FISH ELEVATORS

It is interesting to compare in summarized form the various features, both good and bad, of these two methods of passing adult fish over dams. The following table includes most of the important features.

Comparisons of Some Features of Fish Locks and Fish Elevators

Feature	Fish lock	Fish elevator (trapping and trucking or barging operation)
Capacity	Limited	Much greater
First cost	Less at lower dams under 60 ft (20 m), about equal between 60 and 200 ft (20 and 60 m)	Less at dams over 200 ft (60 m)
Delay at entry	Quite possible	Possible but less likely with good barrier dam or guidance facilities
Delay by repeating cycle	Quite possible	Unlikely
Injury and/or stress	Possible during cycle	Possible during loading transport and release
Operating costs	Low	High
Operating experience	>40 years with Atlantic salmon; five years with various South American species	>40 years with chinook and coho salmon and steelhead trout; >20 years with Caspian drainage fish; >30 years with shad and alewives
Use by downstream migrants	Frequently used, depending on local conditions	Cannot be used

It should be recognized that the effects of fish locks and elevators on the physiology of the fish being handled by existing installations have not been fully evaluated. The long-term effect cannot be determined directly even from the trend in abundance from cycle to cycle, as it is invariably obscured by other effects of the dam at other stages in the life cycle of the fish involved. There is reason to believe, however, that in cases where the other effects are not severe, both methods are successfully maintaining runs of migratory fish that would otherwise have been lost by the construction of the dam.

4.6 LITERATURE CITED

Anonymous, 1950. The Glenfield hydraulic fish elevator, *Br. Eng.,* Feb. 3 pp.

Deelder, C.L., 1958. Modern fishpasses in The Netherlands, *Prog. Fish Cult.,* 20(4), pp. 151–155.

Fish, F.F. and M.G. Hanavan, 1948. A Report upon the Grand Coulee Fish Maintenance Project, 1939–1947, U.S. Fish & Wildlife Serv. Spec. Sci. Rep. No. 4, pp. 151–155.

Hamilton, J.A.R. and F.J. Andrew, 1954. An investigation of the effect of Baker Dam on downstream migrant salmon, *Int. Pac. Salmon Fish. Comm. Bull.,* 6. 73 pp.

Kipper, S.M., 1959. Hydroelectric constructions and fish passing facilities, *Rybn. Khoz.,* 35(6), pp. 15–22.

Klykov, A.A., 1958. An important problem, *Nanka i Zhizu,* 1, p. 79.

Moffett, J.W., 1949. The first four years of king salmon maintenance below Shasta Dam, Sacramento River, California, *Calif. Fish Game,* 35(2), pp. 77–102.

Nemenyi, P., 1941. An Annotated Bibliography of Fishways. Univ. Iowa, Stud. Eng. Bull. No. 23. 72 pp.

Pavlov, D.S., 1989. Structures Assisting the Migrations of Non-Salmonid Fish: U.S.S.R., FAO Fisheries Tech. Pap. No. 308, Food and Agriculture Organization of the United Nations, Rome. 97 pp.

Quiros, R., 1988. Structures Assisting Migrations of Fish Other Than Salmonids: Latin America, FAO-COPESCAL Tech. Doc. No. 5, Food and Agriculture Organization of the United Nations, Rome. 50 pp.

5   FENCES (OR WEIRS) AND BARRIER DAMS

5.1 BARRIERS — GENERAL

History does not record when humans first became aware that fish migrated up and down rivers and could be forced into pens and traps by strategic placement of barriers across the stream. It must have been in prehistoric times, because even today the most primitive peoples have this knowledge, which apparently has been passed down for generation after generation. Indians of the Pacific Coast of North America trapped salmon by this means, and similarly almost every country with rivers supporting migratory fish have records of such traps in their history. Many of the primitive designs are still in use today, as shown in Figure 5.1.

The original traps and barriers were made by lacing together thin branches from trees to form a screen of vertical slats supported on posts driven into the river bed. Where the trapping of fish for commercial use has survived, modern fences and traps made of steel, such as those described by De Angelis (1959) in Italy, bear a marked resemblance to the primitive brushwood fences.

Wood and steel fences of much the same design are now being used for scientific purposes. They have enabled biologists to obtain exact counts of adult fish migrating upstream. With the addition of sufficiently fine-meshed wire screen, they have also enabled the biologists to determine the numbers of young fish migrating downstream. These counts, together with data on the size and condition of the migrants, have contributed greatly to the advancement of scientific knowledge on the life history of migratory fish in rivers.

Fences have also proved to be necessary during the present century in fish cultural operation. When it is desired to capture and hold salmon, trout, and other fish to obtain eggs for hatcheries, the most convenient method has nearly always been the construction of a fence and trap.

These fences, which are now widely used for scientific studies and in fish cultural work, are also known by other names. The name *rack* is often used in the U.S. and is no doubt derived from the term *trash rack* used by engineers to describe the vertical-bar grating placed over water intakes to keep out debris. The word *weir* is also used to describe a fish counting or trapping fence. It is used in this sense more

**Figure 5.1   A primitive barrier for the capture of fish constructed by Indians
on the Klukshu River, Yukon territory.**

commonly in Europe, but it is also used in North America. The word *fence,* however,
is perhaps the most descriptive, is fairly widely used, and for the sake of uniformity
will be used throughout this text.

Because fences are in reality low, pervious dams, it is not surprising that a further
modern development has been the actual consideration of dams to serve similar
purposes to those described for fences. These have been given the name *barrier dams*
or simply *barriers,* because of their use as barriers to migration of fish. They have
come into use only recently, for two specific purposes, and have proved their value
many times. The first use has been to provide a means of guiding fish into a trapping
and holding facility prior to trucking as described in the previous chapter, and the
second is to prevent entry to a lake or pond area that has been cleared of undesirable
species by poisoning or other means.

All these uses, together with examples, will be described in more detail in this
chapter.

5.2 ADULT COUNTING FENCES — SELECTION OF THE SITE

Almost every fisheries biologist who has worked on salmon has come into
contact at one time or another with a fence used for counting adult salmon on their

upstream migration. Many biologists have had to build such structures, and have sometimes been dismayed when the product of much hard labor has disappeared downstream as the result of an unexpected freshet. It should be added, of course, that biologists have not been alone in experiencing such losses. Even well-qualified engineers have lost similar structures as a result of a lack of appreciation of the runoff potential of a river, or more often as a result of inadequate funds, which resulted in an expendable structure.

In designing an adult counting fence, one is faced immediately with two opposing needs. The first is for economy, and the second is for stability. With regard to the first, the exact enumeration of a population of fish is usually only one part of a larger biological study, and it is seldom worth a very large expenditure. A less exact population estimate usually can be made by means of tagging, so that a counting fence is only one of at least two possible alternative methods. On the other hand, the construction of a stable counting fence, operative over the desired range of river discharges, is usually an expensive undertaking. This is particularly true if the river channel is wide, the variation in discharge is large, or the foundation conditions at the proposed counting site are unfavorable. It may be necessary under such conditions to sacrifice a certain degree of stability in favor of economy.

Unfortunately, the need to obtain the count of fish at a certain point on the stream usually sets narrow limits on the available choice of site, and many fences have failed because of poor foundation conditions or low banks that were inundated by the dammed-up river during the first high freshet.

It is important to realize, therefore, that a stable, dependable fence will require careful choice of site and careful design and that it probably will be more expensive than at first anticipated. When these factors are soberly evaluated and considered along with the delay to migration that is a fundamental drawback of every fence, it may often be decided to use some other method of enumeration. However, assuming these factors have all been considered and it is desired to proceed with the installation of a reasonably stable fence, the following remarks may prove useful in selecting a design and estimating the costs.

Choosing the site for a fence is the first step. The design will depend entirely on the site, but to choose a site, certain features of the design must be understood. As pointed out earlier, a fence is a form of extremely pervious dam. The slats in the fence, whether they are of wood or steel, will have to be spaced closer together than the thickness of the smallest fish it is to stop. This means that the smaller the fish to be counted, the smaller the openings and the greater the head loss through the structure. If debris collects on the fence, and it is an extremely rare case when it does not, the openings will be further reduced and the head increased. During a severe freshet, there is likely to be an excessive amount of debris in the river, and a fence can become completely clogged in a very short period of time. This places the full head on it at a time when other effects of high discharge, such as maximum velocities, may be weakening the structure by erosion. Figure 5.2 shows the normal head loss through a fence when it is clean and in good operating condition, and the maximum head when it is blocked completely during a freshet. If it is intended that the fence should be operative during freshets (i.e., the slatted panels left in), it should be designed to withstand the stresses imposed by a complete blockage during maximum

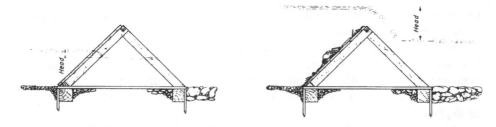

Figure 5.2 Sections through an adult counting fence illustrating the normal operating head (left) and the greatly increased head (right), when the fence becomes clogged with debris during a freshet.

flood flows. It is often too expensive to design a fence to meet this criterion, and fences are therefore frequently designed so that the panels can be removed during freshets, and due allowance is made for any missing biological data that will result.

In this case, one would design the fence only for river stages up to arbitrarily set limits, which in many cases are dictated by the site itself. For example, if a certain point on a river is selected for the construction of a fence, it may be found to have a good compacted gravel or boulder foundation but only low banks. Therefore, while the fence could stand overtopping, it could still fail because the water backed up on its upstream side could run over the banks and erode a new channel around the fence.

In another area selected, the river might have high banks, but both the bed and banks could be composed partly or wholly of sand or some other easily erodible material. In this case, while the river could not overtop the banks, percolation under the structure could quickly undermine it and cause it to fail when high heads were imposed on it.

A good fence site therefore has many of the same qualities as a good damsite. A rock foundation, compacted gravel or boulders are desirable, and high banks of similar materials are necessary. However, one should look for a reach of river as wide as possible.

Perhaps it should be explained further here why choosing a site with the greatest possible river width is important. It can be easily visualized that the wider the river is, the shallower it will be, other factors being equal. Considering the case of a solid dam across a river, the wider the river and the longer the dam, the shallower the depth of flow over the crest, and therefore the lower the head on the dam. In any unit length of the dam, then, the reduced head means less energy to be dissipated, which in turn means less turbulence and as a result less erosion below the dam. The same reasoning applies to a counting fence. The fence should be made as long as the river width at the site will permit, in order to reduce erosion and scour to a minimum.

Assuming that one has found a site with what appear to be the necessary physical features, a rough preliminary check should be made of the hydraulic conditions the fence will have to withstand. It is perhaps too much to expect to have available in most cases a series of gauge readings with corresponding discharges over a full range of flows at the site. It is often possible, however, to find evidence of previous flood levels by means of marks on rocks, trees, or shrubs on the nearby banks, or by finding the level of deposition of sand or silt on the banks. If it is desired to operate the fence above these flood levels, this elevation must be added to the expected head loss through the fence, and the crest of the fence must be built to the resulting level. This would give a fence that would just be overtopped by the highest flood, provided that it could be kept free enough of debris to maintain the head loss within the design limit.

As an example of this preliminary check, it might be found that at a certain site the river cross section averages 1 ft in depth at normal flow, but there is evidence that the depth has reached 3 ft during floods. Allowing for a foot of head loss through the fence, which means that considerable debris could collect on it before it was cleared, and a further foot of freeboard to take care of extra high floods and allow access to the fence at these levels, the fence should then be built to a height of 5 ft above the average stream bed. A check should be made to ensure that the banks are at least this high and preferably much higher. The fence itself must then be designed to provide at least as much unobstructed cross-sectional area of river below this level as existed in the river in its natural state.

Because of the importance of the preliminary site selection for a fence, it might be well to summarize the steps described above, as follows:

1. **Width** — Select the widest stretch of river available, provided that other requirements can be met.
2. **Foundation** — Select the site with the best possible foundation; rock or compacted gravel and boulders are preferred over sand, silt, or pure clay.
3. **Banks** — Make sure the banks are high enough to contain flood levels, including head loss due to the fence.
4. **Flood levels** — Make the best possible preliminary estimates of flood levels or water levels at the highest river stage to which it is desired to operate the fence, add expected head loss due to the fence, and ensure that it can be built to this height or higher if some freeboard is desired.

5.3 ADULT COUNTING FENCES — DETAILED DESIGN

The amount of effort and care put into the detailed design of a counting fence should be proportionate to the cost. For a small fence on a small stream, a satisfactory structure can often be built on the basis of the preliminary check already described. If experienced workpeople are available, it may not even be necessary to draw plans. It is likely, however, that a greater safety factor than is really necessary will be incorporated in the fence design under such circumstances if failure is to be avoided. Some care in design is always desirable, wherever possible.

As the fence cost increases, it becomes more important to protect the investment against loss, and this can be accomplished only by careful design. Furthermore, a larger structure offers more opportunity for economizing by elimination of unnecessary safety factors as a result of careful design based on sufficiently accurate river and site data.

Assuming that the preliminary check has indicated that the site chosen seems feasible, standard engineering procedures can then be followed in as much detail as is compatible with the cost or importance of the structure.

A topographic survey of the area should be reviewed and refined where necessary to ensure that the design is based on the most accurate available data. For larger structures, gauge readings at a wide range of flows are almost essential, and if these can be related to a mass hydrograph of the river recorded at some point not too far distant, so much the better. These data will permit a much better estimate to be made of the range of flows to be expected during the period in which the fence is to be in operation, and to relate this range of flows to actual tailwater levels that will occur below the fence.

The head loss to be expected can be checked more accurately at this stage as well, although in the writer's experience the loss of head from debris collecting on the fence (which can only be determined from experience) is likely to be far greater and therefore will be the determining factor. The head loss by passage of water between the pickets is likely to be only a few inches. It is due largely to contraction of the jets passing between the pickets or bars, so that the head loss can be decreased considerably by rounding the upstream corners of the pickets or in some cases using round pickets or bars. The use of coarse wire screen decreases the head loss when the screen is clean, but increases the tendency to collect debris, which as pointed out, is the governing factor.

The following are some values of the head loss through trash-free fences calculated from an empirical formula that has been used for determining head losses through trash racks. The formula is

$$H = 1.32 \frac{(TV)}{(D)} (\sin A)\left(\sec \frac{15}{8} B \right)$$

where H = head loss in inches
T = thickness of pickets in inches
V = water velocity below fence in ft/sec
A = angle of inclination of pickets with horizontal
B = angle of approach flow with pickets
D = center-to-center spacing of pickets

In the following table the fence is considered to be placed across the stream at right angles to the flow, and the angle B is therefore zero. The table lists some solutions under conditions that are fairly typical for adult counting fences.

Bar thickness (in.)	Spacing center to center (in.)	Angle with horizontal (°)	Velocity (ft/sec)	Head loss (in.)
0.5	1.5	30	4	0.4
		60	4	0.7
		60	10	2.0
1.5	3	30	4	0.7
		60	4	1.1
		60	10	3.0

For a more sophisticated version of this formula, the reader is referred to the U.S. Department of the Interior (1987), which is a complete reference on the design of small dams.

Since the larger part of the total head loss will probably be caused by trash accumulating on the fence, it is well to try to find out what the experience has been at other installations on the same or on a similar river and to use this information as a guide for judging what to allow over and above the head loss calculated by the formula.

When the hydraulic requirements have been thoroughly checked, a preliminary layout can be made on the topographic plan, and a cross section of the river can be drawn showing the fence to the required height. If everything fits satisfactorily, consideration can then be given to the structural design of the fence.

The main elements of the fence are usually (1) an apron across the river, flush with the river bottom, (2) abutments to full flood height on the bank at each end of the apron, and (3) the fence assembly, which is attached to the apron and to the abutments at each end.

The apron is necessary to anchor the fence to the stream bed and to prevent scour and erosion, which would endanger the structure. In the case of timber crib structures, the apron itself is merely a cover of planking over the crib below, and the stability of the structure is derived from the crib rather than the apron. In this case the fence elements should be secured through the apron to the crib members, or directly to the apron if it is designed to be strong enough and fastened securely to the cribs. If a concrete apron is used, it can be designed to be thick enough for the stability to be achieved by its weight alone. Where even a small pile driver is available, it might be desirable to base the structure on piles. In some cases for small structures hand-driven piles have been used in preference to hand excavation and placement of timber cribs. Where piles are used, they can be cut off flush at the top of the stringers spanning between them and the apron can be placed over the top. If a concrete apron is used, it can be poured over the pile tops, provided that suitable attachments are used to take care of uplift and overturning.

It is assumed in the foregoing that the foundation is pervious gravel, in which case a sheet-pile cutoff wall is desirable across the river at both upstream and downstream faces of the apron. The purpose of this cutoff wall is to lengthen the paths of percolation of water through the river bed under the structure as much as possible. As these flow paths lengthen, the resistance to flow increases, and therefore the less chance there is of scour and undermining of the structure. Both the length of the sheet piling and the width of the apron (from upstream edge to downstream edge) will have to be set arbitrarily unless a great deal of detailed study of the foundation materials is made. It is unlikely that such a study would be warranted for a counting fence, so it will not be elaborated further here.

Usually the depth to which the sheet piling is driven is determined by the ease of driving with the available equipment. Where it must be driven by hand, the results may not appear very satisfactory but it is still worth the effort. Where possible, it is suggested that an attempt be made to drive it to a depth about one half the width of the apron.

In the rare case where a suitable rock foundation can be found, it may be possible to eliminate the apron completely, or at the most utilize a very thin layer of concrete to level off the base. The fence can then be attached directly to the rock apron by bolts grouted in holes drilled in the rock or concrete. A fair amount of unevenness in the

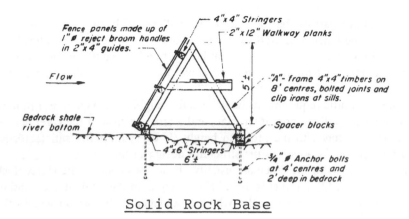

Solid Rock Base

Simple Hand Driven Pile Base

Figure 5.3   Cross sections showing four methods used in constructing adult salmon counting fences on the Pacific Coast of North America.

surface of the apron can be tolerated as long as apertures large enough for the fish to squeeze through are avoided. Figure 5.3 shows cross sections of several fences, including one based on solid rock.

The steel pile and timber crib base fence shown in cross section in the figure is described by Craddock (1958). It will be noted that no apron was used in this installation; this was feasible only because of the rather small variation in flow in the creek (from 50 to 275 cfs). Even then, it was necessary to use a form of cutoff wall of 3- by 6-in. timbers, and a wire screen buried upstream of this. The apron consisted of paving of larger stones downstream of the cutoff wall. It is unlikely that this type of fence would withstand flows much in excess of the stated maximum of 5 cfs/ft of length. The simple hand-driven pile base fence shown in Figure 5.3 is described by Hunter (1954). The apron is composed of 2-in. planks and is 12 ft wide. The planks

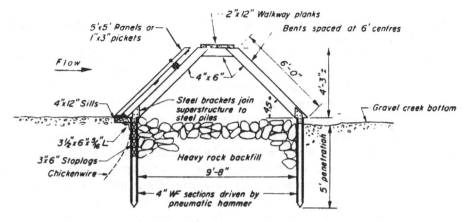

Steel Pile and Timber Crib Base

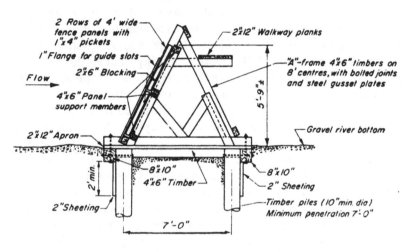

Machine Driven Wood Pile Base

Figure 5.3 (continued).

are fastened to a series of five 4- by 6-in. sills laid across the stream bed in rows, which are in turn fastened to small hand-driven piles. A short, sloping sheet-pile cutoff wall was used, which was really an extension to the upstream edge of the apron. While the steel pile and timber crib fence design referred to above is not too well balanced because it has more secure piling than needed, but lacks an adequate apron, the simple hand-driven pile fence design is just the opposite, having an adequate apron but very weak piling. Flows for the latter are stated to vary between 10 and 300 cfs, which corresponds to only 3.3 cfs/ft of the 90-ft length.

The machine-driven wooden pile fence shown in Figure 5.3 is anchored to two rows of 8-in. diameter piles, spaced at 4-ft centers across the river. It has a somewhat shorter apron than normal because of the strength in the piles and has cutoff walls at both upstream and downstream faces of the apron. This fence has withstood floods of up to 22 cfs/ft of length.

The second main element of an adult counting fence is the abutment at each bank. These should be built integrally with the apron, using the same type of construction where possible as this results in a well-integrated, unified structure. The abutments usually serve two purposes. First, they protect the banks at the fence from erosion due to the higher velocities created by the concentrated head loss at the fence. Second, on a comparatively narrow stream they can add stability to the structure if they are designed as boxes, which, when filled with bank material have considerable weight and hence high resistance to sliding. In this case the apron must be designed to span between the abutments as well as anchored to the foundation material.

The abutments of a pile structure are usually boxes formed of timber sheathing fastened to stringers as shown in Figure 5.3. The stringers are in turn fastened to four long piles, which have been left standing to well above flood height at each corner of the abutment. This box is then filled and capped with timber sheathing to give it some protection from wave action during high river stages.

Where banks are fairly stable and composed of heavy boulders, and the apron has ample strength in itself, the abutments can be simplified. In this case a single wall running along the bank for the width of the apron may be sufficient.

The third element of the adult counting fence, the picket assembly, usually consists of a series of panels fastened at a slope as shown in Figure 5.3. Each panel consists of a number of pickets at the desired spacing fastened together in a unit. The panel units are constructed in a size that permits easy handling, so that they can be removed for repairs. A typical unit is shown in Figure 5.3. The panel supports are designed most conveniently as small A-frames, as shown in Figure 5.3. However, many variations of this design have been developed such as that used by Craddock (1958) and the one described by Hunter (1954). The latter is probably worthy of comment because it utilizes a much flatter slope (30° to horizontal) than is usual for the main picket panels, and has a hinged flap hanging from the upstream legs of the A-frames and resting on the top ends of the pickets. The flatter slope of the main pickets permits easier removal of debris, and the hinged flap prevents fish from jumping over the fence at the higher flows. This can be quite a necessary feature where species such as coho salmon, steelhead, and Atlantic salmon are being enumerated, because of their jumping ability. The hinged flap can be constructed of boards or plywood, and even canvas has been used on occasion.

At adult counting fences it is seldom practicable to have people on duty 24 h/day to count the fish, so that a trap is usually installed in the fence in which fish accumulate, to be counted out at convenient intervals. A typical trap, shown in Figure 5.4, consists of a V-shaped lead from an aperture through the fence to a box formed of pickets. A gate or gates enable the attendant to release the fish after counting them. The box must be large enough to accommodate the maximum numbers of migrating fish arriving at the fence between the times the box is emptied if delay of the fish is to be avoided.

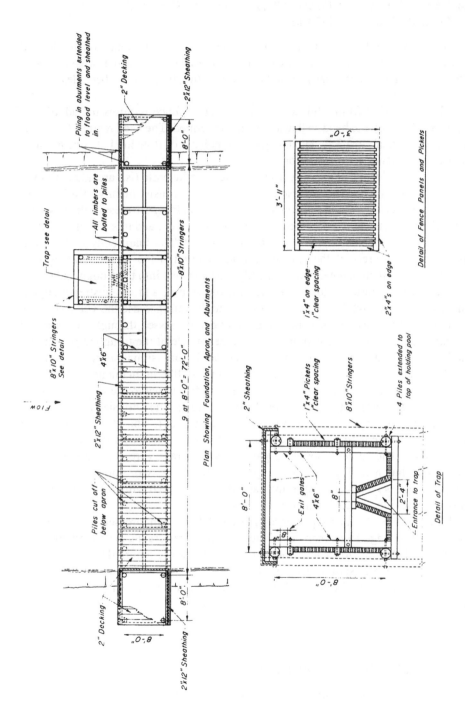

Figure 5.4 Plan of adult counting fence with a machine-driven wood pile base as shown in Figure 5.3, showing details of the trap and fence pickets.

5.4 FENCES FOR ENUMERATION OF DOWNSTREAM MIGRANT JUVENILE FISH

It is more than likely that where the numbers of an adult population are desired, the numbers of the resulting progeny that survive to migrate downstream also are required. For this reason, many recently built counting fences have been designed to serve both purposes.

The conversion from adult to juvenile counting fence requires the addition of stoplogs to build up the head so that a low dam is formed. The head required will vary from at least 1 to almost 3 ft, depending on the volume of flow likely to occur during the period the fence is in operation.

The types of fence commonly used for counting juvenile fish are adaptations of one developed by Ph. Wolf for use in Sweden. They consist mainly of a fine wire screen sloping downward in the downstream direction from the crest of the low dam. A trough is installed at the downstream edge of the screen, and this trough is usually sloped toward either bank where a livebox is placed. The main body of the water passing over the crest of the dam passes down through the screen, leaving only enough on the screen at the downstream edge to carry fish into the trough and thence into the livebox. Various slopes are used for the screen, and the upstream edge is sometimes placed slightly below the crest of the dam to increase the velocity of the falling water and thus increase the amount that can be passed through the screen. A vertical screen is usually placed on the downstream side of the trough to prevent the small fish from jumping out or being washed out by occasional surges of water.

If the main sloping screen is made of 22-gauge (0.028 in. in diameter) wire with 10 meshes to the inch, it will be satisfactory for all fish down to the size range of the smallest Pacific salmon fry, which are about 1 in. long. The wire size and mesh can be increased for larger fish. Coarser screen will of course be stronger and more durable, but the clear spaces between the wires must not be so large that a fish can pass through. The screen should be galvanized or brass wire for durability. The installation shown in Figure 5.5 is typical, except that two sloping screens are arranged in steps from the crest of the dam, the upper one having a coarser mesh to assist in removing as much water as possible.

In the north of Scotland, panels composed of 1- by 2-in. dressed lumber spaced 5/8 in. apart on edge have been substituted for the wire screens described above on installations below the Meig and Luichert Dams. These are reported to be successful for counting Atlantic salmon smolts. It is not known whether any of the smaller migrants pass uncounted through these, but it is considered that small mesh screens are more likely to give a complete count of all migrants, and are preferable where there is a choice.

5.5 BARRIER FENCES FOR ADULTS

It has been pointed out that adult fences have been used for fish cultural purposes as well as for the scientific purpose described earlier. While there is little difference in objective between the two, since they both aim at capturing all the fish migrating upstream, there is a noticeable difference in the type of installation. A fence installed

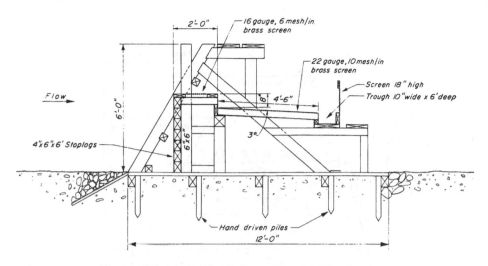

Figure 5.5 A cross section of the simple hand-driven pile base counting fence (as shown in Figure 5.3), with adaptation for counting downstream migrants. (From Hunter, J.G., 1954. *Can. Fish Cult.*, 16, pp. 27–34. With permission of Canada Department of Fisheries and Oceans ©1954.)

for a fish cultural operation is usually considered to be a permanent installation, and is therefore of much more substantial design and construction than the adult counting fences described in this chapter.

In this section we describe a few existing installations to illustrate this difference. As details of the structural design, as in the case of barrier dams in a later section, are beyond the scope of this text, only general design requirements can be discussed.

The timber crib fence, which is shown in the lower view in Figure 5.6, was constructed for the purpose of diverting pink salmon and small numbers of other species out of the main channel into an adjacent artificial spawning channel. Although it is a permanent installation, timber construction was considered adequate with the provision that renewal would be required frequently.

The foundation material was fine, easily eroded gravel, so that the timber box and apron design was decided on; it has withstood subsequent floods satisfactorily. The fence is quite low, because pink salmon are not jumpers and in fact are very weak swimmers. The fence site was in an area where the river had a wide flood plain so it was possible to make the fence quite long, which in turn helped to reduce the flood discharge per unit of length to a value as low as possible. Even so, the apron has been made a total of 19 ft in width, with the downstream part depressed 8 in. below stream bed to act as a shallow spillway bucket to assist in energy dissipation. The walls of the main box or crib under the fence were tightly planked in the open trench excavation, so that they act as sealed cutoff walls, stopping nearly all percolation under the structure to a depth of 5 ft below the creek bed. Further safety features included are the small plank apron extending 3 ft upstream from the structure, and the extension on the downstream edge of the apron consisting of a layer of 300-lb riprap 5 ft wide. The fence also includes a section of similar design except that a solid

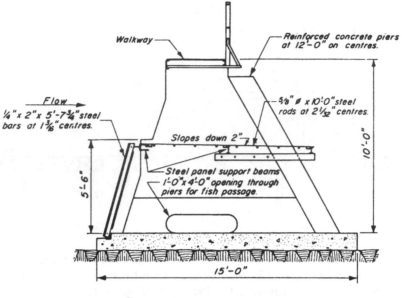

Concrete and Steel Fence

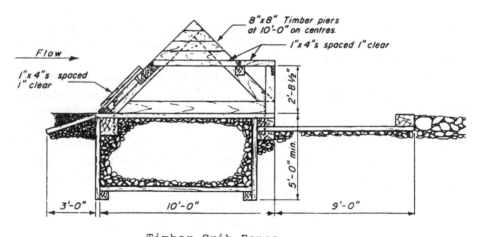

Timber Crib Fence

Figure 5.6 Cross sections showing different methods of construction of two permanent fences used for diverting salmon out of rivers in Canada.

crib is substituted for the picket panels. This section has no overflow during normal flows and permits spills only during floods. The abutments are box cribs of square-sawn timber sheathed outside with 3-in. planking and filled with gravel.

The concrete fence, also shown in Figure 5.6, is constructed of concrete and steel. It was used to divert upstream migrant coho salmon and steelhead into a fishway leading to trapping facilities. The apron is a thin reinforced concrete slab, supporting reinforced concrete piers spaced at 12-ft intervals across the river. The fence panels are supported along the top by steel beams spanning between the piers, and by a structural steel angle at the bottom embedded in the upstream lip of the apron. The fence panels are composed of 1/4- by 2-in. steel bars on edge spaced 1⁵/₁₆ in. apart.

Unique features are the low overall height of the fence and the addition of horizontal panels across the top. This combination results in a less serious cleaning problem than would be the case if a higher fence were used without the horizontal panels. The horizontal panels actually slope slightly downward in the downstream direction, so that most debris is swept off by the current. At the same time, these panels prevent fish from jumping over the fence at higher flows when the water level downstream is approaching the height of the fence. When flood flows overtop this fence, as they do occasionally, some fish are able to pass over, but this occurs only rarely. There are few fish ascending at the time of year when these floods occur, and velocities then are so high that only exceptionally powerful swimmers are able to surmount the fence. The design is therefore considered to be a practical one for this location under the existing conditions.

Barrier fences have been used in other countries to trap fish for fish cultural purposes. Figure 5.7 shows two fences installed for this purpose at Loch NuCroie and Loch Poulary in the north of Scotland. These are constructed of concrete on gravel foundations. Concrete piers spanned by steel beams support the fence panels. The panel construction, bar spacing, etc., are similar to those already described, except that there are no horizontal panels. The alignment of the fences in a "V" pattern in plan to assist in leading the fish to the traps, which project upstream, will be noted. It will also be noted that only one trap is at the apex of the "V" in each case, the others being distributed across the fence. These fences are used to capture entire runs of Atlantic salmon for maturing and stripping, and hatchery rearing of the progeny.

5.6 COSTS OF FENCES

From an analysis of the costs of many fences constructed for Pacific salmon diversion over the years, the following general observations can be made:

1. The unit cost of a fence can be stated as the cost per cubic foot per second or per cubic meter per second of flood flow to be handled. In general, this unit cost tends to decrease as the size of the flood flow increases. The range is roughly from \$60/cfs for small fences with a low flood flow to \$20/cfs for large fences on good foundation and a high flood flow. The metric equivalent is \$2100/m³/s for the small fence to \$700/m³/s for the large. (Prices are in 1987 U.S. dollars.)
2. There is a wide variation in total size, foundation materials, flood flows, and total length of fence, so that the figures on unit costs given here must be used with caution. Any specific deviation must be accounted for in considering a particular installation.

**Figure 5.7   Two fences for trapping Atlantic salmon in the north of Scotland:
the fences at Loch NuCroie (top) and at Loch Poulary (bottom).**

5.7 BARRIER DAMS

The problems encountered in locating the entrance to a fishway below a dam so
that adult fish migrating upstream will find and enter it with the least delay were
described in an earlier section. It will be readily appreciated that there will be many
dams where ideal entrance conditions will be extremely expensive or even physically
impossible to attain. In searching for an answer to these difficult cases, the idea of
building a second dam downstream, which could be quite low, but designed to

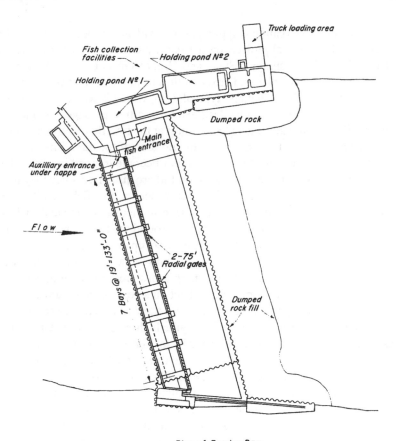

Plan of Barrier Dam

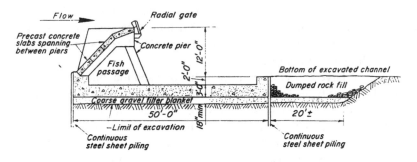

Typical Cross Section

Figure 5.8 **Plan and cross section of the barrier dam constructed on the Baker River in Washington State as part of the collection works for Pacific salmon.**

incorporate entrance conditions approaching the ideal, has been evolved. This scheme has been used on several projects on the Pacific Coast in North America, including the Baker River in Washington State, and Great Central Lake on Vancouver Island.

Figure 5.8 shows a plan and cross section of the barrier dam on the Baker River near Concrete, WA. It is a buttress type of dam, which is well suited for use as a fish barrier. This is because it can be designed with a free passageway under the nappe of the overflow; fish that are tempted to jump at the overflow and penetrate it fall into this passageway rather than impinging on the face of the dam. Therefore, at periods of low flow, fish approaching the dam cannot become trapped in any of the spillway bays and remain there to injure themselves, but are quite free, once they penetrate the spill, to move to the side to find the entrance to the fish facilities. Since this installation was completed, runs of up to several thousand coho and sockeye salmon annually have been trapped and transported around the two dams of the Puget Sound Power and Light Company, which are located upstream. Each of these two main dams is approximately 300 ft high, and the single barrier dam and trucking facility is able to serve both, because all spawning areas are above the upper dam.

It will be noted that the barrier dam is built on a pervious foundation, with steel sheet-pile cutoff walls at the upstream and downstream edges of the apron. The cost of the dam is not known, but it is believed to have been quite high because of the type of construction necessitated by the foundation materials and by the functional nature of the dam.

The other barrier dam installation noted previously is located below a storage dam at the outlet of Great Central Lake on Vancouver Island, B.C. Both dams and the fishway that serves them are shown in plan and section in Figure 5.9. The barrier dam was used here to provide an effective but economical lead to the fishway entrance. A rock outcrop created a natural falls in the general area now occupied by the barrier dam. This rock barrier was excavated and trimmed to the dimensions shown, and a reinforced concrete lip was added to the crest to provide the required height for the barrier. A minimum vertical drop in water-surface elevation of 10 ft, created for all river stages, has been successful in preventing ascent by the salmon. The diagonal barrier leads fish naturally to the fishway entrance. The fishway has vertical slot baffles and pools 8 ft wide by 10 ft long. The dual entrance carries the fishway flow plus auxiliary water taken directly from the forebay and introduced into the entrance through a floor diffuser.

The costs of large barrier dams such as this one and that on the Baker River are not susceptible to the type of unit analysis given for the fences in the previous section. The costs depend entirely on the site, construction materials, location, etc., and can be judged only by taking all these factors into consideration. No rule-of-thumb method for arriving at preliminary costs can be suggested, therefore. In making an estimate, a preliminary design should be prepared, and preliminary costs obtained by calculating quantities and taking all factors into consideration. For structures of this size, competent, and specialized engineering advice is a prerequisite.

At the beginning of this chapter, it was noted that barrier dams have another use: to prevent entry of unwanted species into a lake or impoundment after the unwanted species have been eliminated and the facility stocked with the desired species. Sport fishing demands in North America have dictated this approach. The barrier usually consists of a low weir of timber crib or other design. It is normally not more than 3 ft

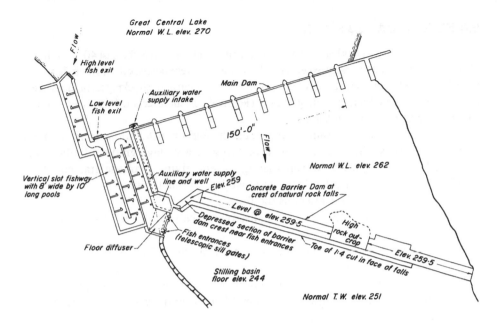

Plan of Barrier Dam and Main Dam

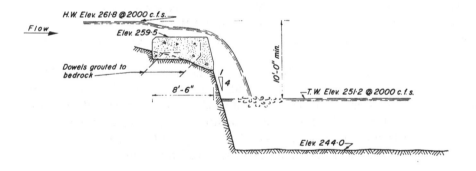

Typical Section Through Barrier Dam

Figure 5.9 Plan of the Great Central Lake Dam and Fishway on Vancouver Island, B.C., and cross section through the modified natural barrier dam formed by the falls over the rock outcrop downstream.

(1.0 m) in height, from spillway crest to apron. A near-horizontal set of wooden slats are placed from the spillway crest sloping slightly downstream over the apron, so as to carry as much of the trash over the dam as possible and to prevent jumping of fish over the crest. The spacing of the slats can be varied according to the type and size of fish it is necessary to prevent from migrating upstream. This type of barrier can be constructed very economically.

5.8 ELECTRICAL BARRIERS

This type of barrier has been experimented with extensively over the past 40 years. In my 1961 edition, two of these early experimental installations were described, and their disadvantages listed. With progress in the development of fish-stunning and -shocking techniques, there has been an increased understanding of how electrical current affects fish in water, and this has been applied in newer installations.

For upstream migrant salmon and trout, the earlier installations utilized either direct current pulsating 10 times per second to give a voltage gradient in the water of about 2 V/in., or a 110-V 60-c alternating current, giving a voltage gradient of 0.3 to 0.7 V/in. It is noted that the latter was installed in much higher water velocities, thus having about the same effect as the former.

Even without thorough testing, the disadvantages were listed as follows:

1. An unknown loss of adults downstream from the effects of shocking and possible predation
2. The cost of bringing sufficient power on site to operate the facility
3. The necessity of choosing a site or making one with evenly graded bottom and sides and maintaining it

These disadvantages, to some degree, offset the comparatively low cost of the barriers at that time.

Since these early installations, government agencies have largely given up experimenting with this type of barrier, leaving it to commercial companies to develop. To date, at least one company (Smith-Root, Inc., of Vancouver, WA) has continued testing and evolved a barrier that apparently goes a long way toward solving the first disadvantage listed above, while still having the other two disadvantages. Called the *graduated field fish barrier,* it is shown diagrammatically in Figure 5.10.

Each of the seven Model GFFB-1.5 U pulse generators provides pulse widths of 8 to 48 ms at rates of 1 to 3 pulses per second. They are typically spaced 1 m apart, and when synchronized, the outputs are additive along the array. Thus they provide a gradually increasing field intensity from downstream to upstream of the barrier, and the field extends to the water surface. The "insulating substrate," which is a special mix of concrete laid across the bottom of the stream, further enhances the electrical field in the water.

This type of facility appears to offer possibilities in certain situations where the site configuration is right and a power supply of 240 VAC is available.

Only a few installations have been made to date. Since they have not been scientifically evaluated, they must be described as experimental until wider usage has proven their worth.

5.9 THE GREAT LAKES LAMPREY BARRIER

A special barrier has been developed for use on the Great Lakes in North America, where the invasion of sea lampreys has been a problem since the completion of the St. Lawrence Seaway. Lamprey control has involved the development of

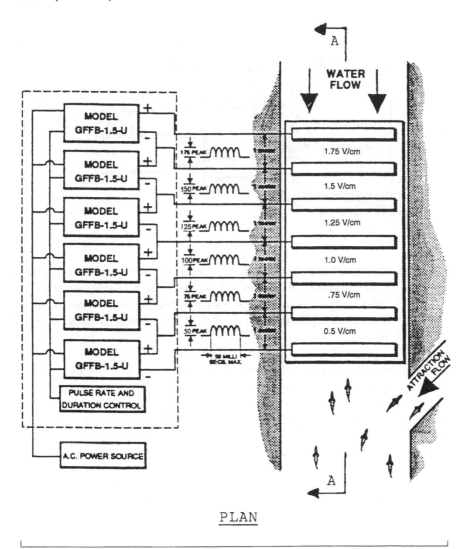

PLAN

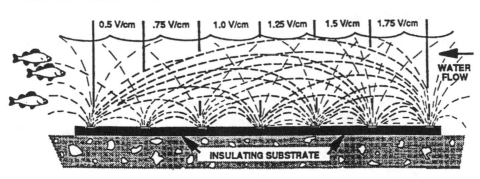

SECTION A-A

Figure 5.10 A graduated field electric barrier as developed and marketed by Smith-Root, Inc., Vancouver, WA 98686. With permission.

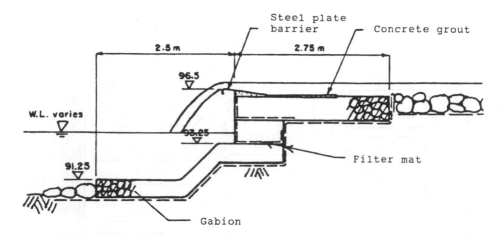

Figure 5.11 A typical lamprey barrier as used by the Great Lakes Fishery Commission on the Great Lakes. The Canada Department of Fisheries and Oceans and the Ministry of Natural Resources of Ontario were the cooperating agents of the Commission. Dimensions are in meters.

this barrier, which must stop the lampreys from ascending tributaries of the lakes to spawn, while allowing trout and salmon to ascend without hindrance.

The barrier development is simply a concrete weir no more than 1 m in height with a steel plate lip extending 23 cm (9 in.) from the downstream face of the weir, as shown in Figure 5.11. The lampreys are faced with a head of about 60 cm (2 ft) to ascend over the weir, and since they are unable to jump, they cannot surmount it. A box trap is usually provided in the weir, which makes it possible to count and dispose of them. The trout and salmon are able to jump this obstruction with ease and proceed upstream. This weir has been found to be very effective in controlling the numbers of lamprey in this area.

5.10   LITERATURE CITED

Craddock, D.R., 1958. Construction of a two-way weir for the enumeration of salmon migrants, *Prog. Fish Cult.,* 29(1), pp. 33–37.

De Angelis, R., 1959. Fishing installations in Saline Lagoons, Gen. Fish. Counc. Medit. Stud. Rev. No. 7. 16 pp.

Hunter, J.G., 1954. A weir for adult and fry salmon effective under conditions of extremely variable runoff, *Can. Fish Cult.,* 16, pp. 27–34.

U.S. Department of the Interior, 1987. Design of Small Dams, 3rd ed., Water Resources Tech. Publ., Washington, D.C.

6   PROTECTION FOR DOWNSTREAM MIGRANTS

6.1 THE PROBLEM

The early chapters of this text deal with facilities to enable fish to migrate upstream over dams and waterfalls. We now turn to the methods of ensuring that the young fish (in the case of anadromous fish) and the grown fish (in the case of catadromous fish) can descend the same path as safely as possible. In the case of anadromous fish, some species descend the river again after spawning, and they too must be considered.

The catadromous fish are mainly eels, freshwater mullet of southern Africa, golden perch, barramundi perch, and Australian bass. These fish are for the most part faced with low dams in Europe, Africa, and Australia, so that there is no great problem involved in their traveling downstream when they mature. Downstream passage over low weirs does not present a hazard. Some eels in Europe, North America, and New Zealand have to pass over high dams, and perhaps through turbines associated with these dams, but to date little study has been devoted to them, except incidentally to other studies. If there is a problem, it will have to await better definition and more research in the future.

The anadromous species on the other hand have been known to face a real problem for 60 to 70 years, and study has been escalated in the past 20 years to try to find a solution. These studies have been confined mainly to North America and to a lesser extent to Europe. They have been wide ranging, but uncoordinated, and devoted mainly to specific problems in specific areas. Only recently have two North American efforts been made to try to define the whole problem and suggest avenues of research.

The first of these has been a modest effort by the Canadian Electrical Association in 1984, entitled *Fish Diversionary Techniques for Hydroelectric Turbine Intakes*. It was prepared by Montreal Engineering Co. with C.P. Ruggles and R. Hutt as the Principal Investigators. As the title suggests, its purpose is to establish if there is a need in Canada for research into this field at this time, and it concludes that research is not presently needed for hydroelectric intakes. One may question this conclusion. But the report does, however, review the diversionary facilities presently in place or under investigation for all kinds of water uses, and suggests where the best prospects lie to improve them.

The second of these is a massive effort by Stone & Webster Engineering Corp. (1986), commissioned by the Electric Power Research Institute, Palo Alto, CA, entitled *Assessment of Downstream Migrant Fish Protection Technologies for Hydroelectric Application*. While the main thrust of the report is to look for a solution for hydroelectric plant intakes, it does review, in a great deal of detail, every type of diversionary device in use and tried experimentally in North America, and some in Europe. These range from electrical screens to submerged traveling screens. They are subdivided into groups as follows:

1. **Behavioral barriers** — Rather than forming a physical barrier, these involve a behavioral response on the part of the fish such as the reaction to electrical screens, air bubbles, hanging chains, lights of various types, water jet curtains, etc. In general these types involve little maintenance because the water passes through freely and there is little or no accumulation of debris. On the other hand, none of them is very promising and most of the installations described have been experimental only.

2. **Physical barriers** — These are complete barriers, although there can be some penetration of them by fish. In addition to penetration, some fish can be killed by barriers in this category by becoming stuck in the mesh or by exhaustion. They include bar racks and fixed screens, traveling screens, drum screens, cylindrical wedge wire screens, and barrier nets. They are subject to clogging, and have a high maintenance cost as well as initial cost.

3. **Fish-diversion devices** — To a certain extent these duplicate the physical barriers, because they include drum screens and submerged traveling screens with the added provision that they are angled to guide fish approaching them. But they also include inclined plane screens and louvers, which will be described later. Some of these have high maintenance costs and some have low costs. They include some of the most promising methods.

4. **Fish-collection devices** — These include the "gulpers" and other forms of collection device used above dams. Operational costs of these are high, although initial costs are comparatively low. The most promising of these will be described later in this chapter.

Obviously there is a consciousness of the problem on the part of hydroelectric owners, as well as owners of intakes of all other kinds. Included in the latter are the irrigation intakes, cooling water intakes for thermal electric stations and other industrial uses and public water supply intakes. Most of these are terminal for any fish migrating downstream, that is, 100% of fish caught in irrigation works or industrial intakes are lost. But fish passing into hydroelectric intakes are not. In fact, a good percentage of them survive.

6.2 FISH LOSSES DUE TO HYDROELECTRIC PLANTS

The losses due to hydroelectric plants vary according to the route taken by the fish on their journey downstream. Losses over spillways are different than those through turbines. In some cases they can be higher, but in most cases they are less.

Losses over spillways have been studied by a number of investigators, and have been summarized by Bell and DeLacey (1972) and by Ruggles (1980). Summarizing them briefly, losses at the dams studied ranged from 0.2 to 99%. In general, losses over the lower spillways were less, while some of the highest spillways had the highest losses. This led to some experiments with air drops from a helicopter, in which Bell and DeLacey found that fish in the range of 10 to 13 cm in length reached a terminal velocity in free fall of 16 m/sec in falls of approximately 30.5 m. Fish in the 60-cm range had terminal velocities in excess of 58 m/sec. The survival of fish of 15 to 18 cm in length was in the 98% range for drops of 30.5 to 91.5 m. Thus it was obvious that for free falls, small fish could survive a drop equivalent to most dam heights, while larger fish would have difficulty in surviving.

However, the height of the dam, while eliminated as an important factor in itself in determining mortality for small fish, was important when combined with other factors. These other factors were abrasion against the spillway surface, rapid pressure changes, and shearing effects (a rapid change in direction of flow). An additional factor was nitrogen supersaturation owing to the deep plunging action of the spill. Some of these effects could be mitigated by designing ski-jump type spillways, provided that a deep enough pool was available to cushion the final drop into the water. These are possible only at high dams though. Another possibility was to design a spillway deflector that directed the flow horizontally into the stilling basin, thus avoiding the supersaturation problem pointed out above, which proved harmful to fish. Both of these types of spillway, and others, are illustrated in Figure 6.1. Another possibility is to design the spillway so as to minimize abrasion and impingement of fish on rocks and concrete. It is obvious, however, that no general set of rules can be formulated for spillway design to optimize fish survival, and each case has to be looked at individually.

Losses in passage through turbines have also been studied extensively. Ruggles (1980) and Bell (1981) summarize these studies for the North American continent and parts of Europe; Larinier (1987) adds some further European results. It is obvious, from these summaries and the studies on which they were based, that the questions are very complex and no easy solutions are available.

The causes of mortality in passage through turbines are listed by Ruggles (1980) as follows:

1. Mechanical damage, due to contact with fixed or moving equipment
2. Pressure-induced damage, due to exposure to low-pressure conditions within the turbine
3. Shearing action damage, due to passage through areas of extreme turbulence or boundary conditions
4. Cavitation damage, due to exposure to regions of partial vacuum

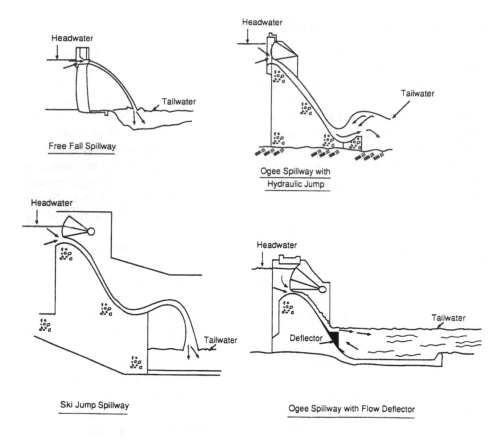

Figure 6.1 Some types of spillways mentioned in text.

Although each of these general categories produced some specific identifiable injuries, many of the injuries could not be associated with one cause alone. Since the type of injury generally differed with the type of turbine, they will be discussed separately in what follows.

There are two main types of turbines in common use at the present time on streams and rivers supporting migratory fish population. Their type names are derived in part from their method of developing power. The mixed-flow type (generally known as a Francis turbine) takes the flow and changes it by 90° in passage through the control gates and turbine. The axial-flow type (generally known as a Kaplan turbine) allows the water to pass through without the abrupt change in direction and develops power through a propeller turbine. The names Francis and Kaplan are derived from the early designers who first developed them. The Francis type is usually used for higher heads, where the head is reasonably constant but the flow may vary considerably. The Kaplan is used for lower heads, where there is relatively large flow and variable head. Both types of turbine runner are shown in Figure 6.2.

For the Francis turbine, the flow, or amount of water passed through the turbine, is controlled by wicket gates at the entrance to the turbine, which can be opened or closed by the operator. It has very closely spaced openings in both the wicket gates,

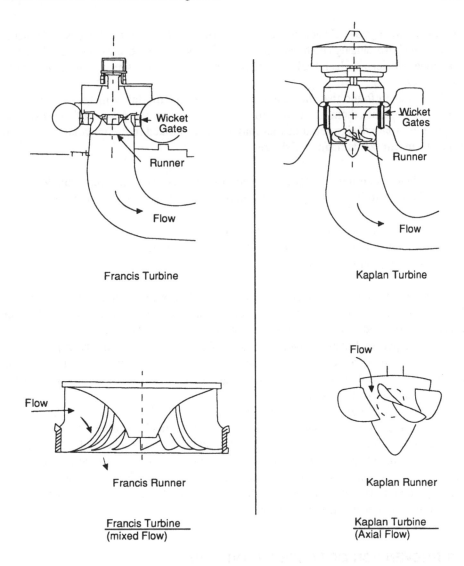

Figure 6.2   The two main types of turbines in use on fish-producing rivers. The differences in clearance of blades will be noted as well as the proximity of the wicket gates to the runner in the Francis turbine.

and between the gates and the turbine blades. For the Kaplan turbine, on the other hand, the flow is controlled by adjustable blades on the runner itself; clearances between the blades, and between the blades and the housing are generally much larger. While this is generally the case for each type, variations do occur among turbines to suit particular conditions.

For Francis turbines, small clearance between wicket gates and leading edges of the runner have contributed to much of the loss from mechanical damage as described by Ruggles (1980) under item 1 above. In addition, the higher runner speed has contributed to these losses. Kaplan turbines have fewer blades (ordinarily 6 to 8), they are more widely spaced, and they operate at slower speeds, so the fish are less likely to be injured mechanically. Kaplan turbines, however, have a greater chance of causing the type of injury listed by Ruggles under items 2, 3, and 4 above, although both types can cause damage through being operated at less than maximum efficiency, which leads to negative pressures, shearing stresses, and cavitation of the runners.

It has been found that the best way to reduce fish mortality in existing installations is to operate them at maximum efficiency during the period of downstream migration. Some design features that may be incorporated in the installation of both types to minimize damage to fish are

1. Have maximum clearance between the wicket gate and runner blades.
2. Have a turbine setting as deep as possible relative to expected tailwater levels.
3. Have a minimum turbinc blade speed.
4. Design the turbine to operate at maximum efficiency. (This is usually a design goal in any case.)

Operating a turbine at maximum efficiency during the period of migration can have a very beneficial effect on the degree of fish mortality, and survival of the fish might be expected in the 90 to 98% range under these conditions.

It is suspected, but not proven, that larger fish suffer comparatively larger losses. The fish tests summarized here were conducted with fish mostly of a length of 15 cm (6 in.) or less, and it is expected that larger fish would suffer more mechanical damage because of their size.

Losses at hydroelectric plants are one of the most vexing problems facing fishery biologists and engineers today. Attempts have been made to divert the fish, with various barriers or guides, some successfully and others less successfully as we shall see later in this chapter. Various types of screens have been developed, some only partially successful. There have been attempts to only partially divert the fish, with the hope that a 10% loss can be reduced to, say, 5%. All these will be examined in the section on screens later in this chapter.

6.3 PREVENTION OF LOSSES IN INTAKES

It is convenient to discuss this topic under the headings listed in Section 6.1 (with slight alterations for convenience).

6.3.1 Behavioral Barriers

In an effort to avoid the disadvantages of complete physical barriers and their associated problems of debris accumulation and subsequent clogging, every conceivable alternative means of causing fish to move away from intakes has been tried. Most of these have been only tried experimentally, but some have been field tested. Curtains of air bubbles, hanging chains, flashing lights of various types including strobe lights, and various types of sound barriers have been tested with results that were not very promising, with the possible exception of sound barriers under certain conditions.

Electrical barriers have also been tested extensively, and this is one of the few types to show any promise of even limited success. The trouble with this type of barrier at downstream intakes is the direction of flow. At upstream barriers of this type (see Chapter 5) the fish either are repelled or go farther into the electrical field until they are partly narcotized, and are carried out by the current to recover and try again. At downstream barriers, on the other hand, fish that enter the electrified field too deeply become narcotized and are carried by the current even deeper into the electrical field, from which there is little hope of escape.

In general, the behavioral barrier may have the best chance of application to lake intakes for cooling or domestic water, where there is no perceptible velocity at some distance from the intake and the barrier can be placed at such a point.

6.3.2 Physical Barriers and Diversion Devices

These are the most commonly used barriers, and have been developed over the years in many different forms. The two types have been combined in this section as they overlap. For every barrier to fish migrating downstream there must be an escape route. In other words, it is necessary to provide a *bypass* as well as a screen, and often the effectiveness of the screen depends on the success of the bypass. Figure 6.3 shows diagrammatically a typical intake on a stream or diversion from a stream, complete with bypass to offer the fish a safe route to continue their journey downstream. In this case the diversion is from a river; 10 cfs is withdrawn from the river, nearly all of which is screened, with only 0.25 cfs being returned to the river through the bypass. In designing such an installation, several questions will arise, such as how wide and deep the intake conduit should be, what mesh size and wire the screen should be, how large the bypass opening should be, and where it should be placed. The answers to these questions depend entirely on the swimming ability, physical size, and behavior of the fish, and it is proposed therefore to discuss these problems separately in the sections that follow. The first of these sections deals with the screen area, size, and shape of approach channel, and the velocity of the water approaching the screen. The next section covers the problem of mesh size, which is determined by the physical size of the fish to be screened. The last section discusses the bypass. Its requirements are based largely on the fish behavior and, to a lesser extent, their swimming ability.

6.4 APPROACH VELOCITY AS DETERMINED BY SWIMMING ABILITY

Many biologists have conducted experiments to determine the swimming capabilities of fish. Some of these experiments have been conducted on small fish in the size range one might wish to keep out of water intakes. Kerr (1953) reports on tests of the swimming speed of striped bass and chinook salmon in the size range from 1 to 7 in. in length. Brett et al. (1958) describe tests on sockeye and coho salmon from about 2 to 6 in. in length. Bainbridge (1960) describes tests on dace, trout, and goldfish ranging from 2½ to 11 in. in length. The data from these experiments is most useful in solving screening problems and particularly in determining the approach channel velocities to use in the design.

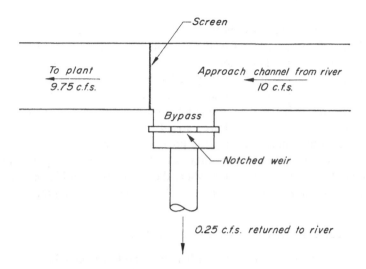

Figure 6.3   **Diagrammatic representation (in plan) of a screened water intake on a river with a bypass to carry fish back to the river.**

While approach conditions vary with screen installations, the hypothetical case shown in Figure 6.3 is probably typical of the most difficult conditions to be encountered. Here it is assumed that the approach channel is a considerable length, such as 100 ft or more. In other words, the screen is far enough from the river that it is unlikely fish migrating downstream and entering the approach channel would return back up the channel to the river after reaching the screen. It is therefore necessary to provide a bypass so that they might continue their journey downstream. However, it is not always possible to design the bypass so efficiently that the fish can find it quickly. While much is now being learned about bypass design, it might be expected that at least some of the fish will remain in front of the screen seeking an exit for an indefinite length of time. If velocities are high, it is improbable that many fish could be delayed more than a day in this fashion and still survive because they would eventually tire and be killed against the screen. So for practical purposes it might be hypothesized that the speed of the water against which they must swim in the area immediately upstream from the screen must be in the same order of magnitude as the cruising speed the fish can maintain for some such period.

Brett et al. (1958) define the cruising speed for sockeye and coho salmon as determined in their experiments as "the swimming speed which a fish can maintain consistently for a minimum period of one hour under strong stimulus without gross variation in performance." While this does not meet our requirement for a speed that a fish can maintain for as long as a day, it is believed it is in the general range we are considering. Extrapolation of Brett's curve, as shown in Figure 6.4a for mean cruising speed at 10°C, gives a speed of 0.5 ft/sec for a mean fork length of 2.54 cm, or 1 in., which is probably a little less than the length of the smallest migrating salmon fry. While the curve for 5°C gives a lower cruising speed, temperatures this low during downstream migration are not common, and we will disregard them for the moment.

Bainbridge's experiments were lengthy, and only one of the graphs, showing the calculated relationship between speed and length of time it can be sustained is included in

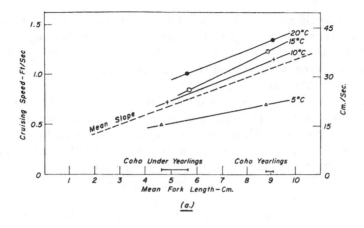

(a.)

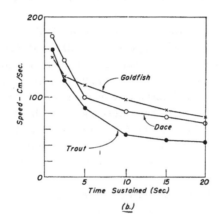

(b.)

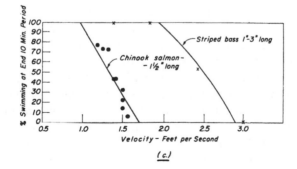

(c.)

Figure 6.4   (a) Mean cruising speed for underyearling and yearling coho salmon, for four levels of acclimation. (From Brett, J.R., M. Hollands, and D.F. Alderdice, 1958. *J. Fish Res. Bd.*, 15[4], pp. 587–605. ©Canada DFD.) (b) Calculated relationship between speed and length of time it can be sustained for three 15.0-cm-long fish. (From Bainbridge, R., 1960. *J. Exp. Biol.*, 37[1], pp. 125–153. With permission.) (c) 10-min velocity endurance curve for chinook salmon 1½ in. long and for striped bass 1 to 3 in. long. (From Kerr, J.E., 1953. Calif. Fish & Game, Fish. Bull. No. 92. 66 pp.)

Figure 6.4b. It would appear to give values for cruising speeds much lower than Brett's, although a direct comparison is impossible because the species are different, and the longest cruising period investigated is only 20 sec. For this period Bainbridge suggests a general relationship of four times the fish's length per second for the speed that can be maintained. This works out to only 0.33 ft/sec for 1-in.-long fry.

Kerr's results are more in accord with Brett's, and since he utilized Pacific salmon in addition to striped bass, his observations are considered to be more applicable to screening problems involving salmon. For striped bass, Kerr reports that 100% of fish averaging 1 in. long maintained a speed of 0.83 ft/sec for 10 min. For chinook salmon averaging 1.5 in. in length, he reports that at velocities of 1 ft/sec, only 92% of the fish tested were still swimming at the end of 10 min. He does not give figures for his tests at lower speeds, but his graph shown in Figure 6.4c would indicate that at velocities of 0.5 ft/sec, 100% of the fry this size could swim for 10 min. He also states there was little difference in ability between bass and salmon of the same length.

It is concluded from these admittedly limited data that for practical purposes the smallest Pacific salmon fry might be expected to cruise at speeds of 0.5 ft/sec for the longer intervals necessary to keep themselves away from screens and to find a safe bypass. Because there is usually a lack of complete uniformity in the velocity of the water approaching a screen, and variations in temperature are also possible, it is considered desirable to allow a further safety factor in the overall design, and an approach velocity of 0.4 ft/sec (12 cm/s) is recommended for screens that are properly maintained in a clean condition. It has been found that as velocities of approach to the screens increase above 0.4 ft/sec, the smallest salmon fry rapidly become exhausted and are swept against the screen and killed. Fry of this size are present in the river only during the spring immediately after emergence from the gravel, and as the season progresses and fry increase in length, it is found that they can withstand progressively higher velocities.

Where salmon smolts must be screened, and no fry are present, the standard quoted can be safely increased. For fish in the length range of 8 cm or more, a recommended safe standard is 1.0 ft/sec. Brett's data supports this conclusion, and it has also been found by experience to be adequate.

As pointed out earlier, several factors can influence application of these standards, and these factors should be evaluated before applying the standards. The two main factors are (1) the possibility of existence of abnormal temperatures and (2) the application to species other than those discussed. With respect to the first, Brett's data give some quantitative indication of the influence of temperature and can be used as a guide to the effect of this variable. With respect to the second, however, experiments on swimming performance of fish have not been extensive, and designers wishing to provide screens for species not already tested may have to resort to performing their own experiments beforehand. Bainbridge's data show the possible range of variation in swimming ability between species for the conditions prevailing in his tests. The results of Kerr's experiments, however, would tend to indicate that for fish about 1 in. in length, there is apparently little difference in swimming ability between different species in the same environment.

When the magnitude of the approach velocity has been decided, the dimensions of the approach channel can be calculated. The shape of the channel cross section is not too important, as long as the cross-sectional area below the lowest water level likely to occur during migration is adequate to produce a mean velocity equal to the approach velocity decided on. Extremes in width or depth should be avoided, however, because of the possible effect on velocity distribution through the screen. Many intakes consist of a channel leading from the river to a pump well, with the screen installed across the channel where it joins the pump well. The suction pipe from the pump is suspended in the well a short distance behind the screen. It is natural, then, that the highest velocities through the screen will be in the area immediately adjacent to the end of the suction pipe, and that lower velocities will occur through the more distant parts of the screen. If the approach channel to the screen is not excessively shallow or excessively narrow and deep, and the suction pipe is a reasonable distance behind the screen, the velocity distribution across the face of the screen will be fairly uniform. In some cases, however, a "Y" or "T" connection on the end of the suction pipe might give a more even velocity distribution. It is preferable, however, if at all possible, to make the channel dimensions such that this arrangement is not necessary.

It is assumed that the surface area of the screen is, for practical purposes, the same as the cross-sectional area of the water in the approach channel in the foregoing discussion. If they are not the same, the smaller one governs and must meet the velocity requirements specified. In other words, if the velocity specified is 0.4 ft/sec, then there must be a minimum of 2.5 ft^2 of screen area for each cubic foot per second of flow. The water in the approach channel then must have the same or larger area of cross section than the screen area.

6.5 MESH TYPES AND SIZES FOR SCREENS

Four types of screen have been used to exclude fish from intakes. These are a simple bar screen, a woven mesh screen, a perforated plate, and finally an improved welded mesh screen known as a wedge wire screen.

The simple bar screen consists of vertical slats or bars like a trash rack, spaced sufficiently close to prevent the fish from entering. It has been useful primarily for preventing entry of larger fish, because the close spacing for smaller fish led to problems of intake fouling by debris and algae; in addition, the narrow spacing restricted the flow into the intake unacceptably. For these reasons simple bar screens have been largely abandoned for modern intakes.

The woven mesh screen is usually made from wire, and it has square openings between meshes. For fixed screens such as the barrier net screen shown in Figure 6.5, a mesh net of rope or synthetic twine has been used. For this special case of an intake on a lakeshore it has been found to be cheap to install and maintain. It should be far enough from the intake for the velocities through it to be well below 0.4 ft/sec (12 cm sec). Experience has shown that the sizes of mesh and of mesh material have to be adapted to the species to be excluded and to the conditions obtaining in the lake or river such as temperature, currents, etc. For an intake on a flowing stream or canal, a wire mesh screen is normally used.

The size of mesh openings required can be simply determined by the dimensions of the fish to be excluded, measured at the deepest part of the body. It is not altogether

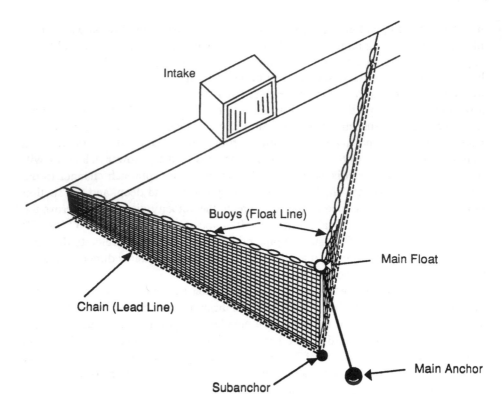

Figure 6.5 Schematic plan of a typical barrier net screen at a lake intake.

satisfactory to decide on the mesh size on this basis, however, because fish bodies are flexible and they can pass through smaller openings than the dimensions would indicate. It is preferable to conduct tests with live fish of the species and size it is desired to screen, using screens with various sizes of mesh openings, and from these determine what mesh size is necessary.

From tests of this kind, and experience, it has been found that wire screen with square openings of the following sizes are satisfactory for preventing passage of fish of the general size and species shown:

Size and species	Size of mesh opening (1 side of square)
Pink salmon fry immediately after emergence from gravel (~1 in. or 2.54 cm long)	0.10 in. (2.5 mm)
Sockeye salmon fry, chinook salmon fry, coho salmon fry (~1.2 in. or 3.05 cm long)	0.12 in. (3.0 mm)
Larger chinook and coho salmon fry (~2 in. or 5.0 cm long)	0.15 in. (3.8 mm)
Yearling salmon or smolts (~3.5 to 6 in. or 8.4 to 15 cm long)	0.25 in. (6.3 mm)

It is usually possible to find a combination of wire size (diameter) and number of meshes per inch that yields about 50% clear open space through the screen. This percentage opening is essential if, as is normally the case, the specified maximum approach velocities will occur. If the percentage of open space is too low, velocities through the screen will be too high and some fish may be forced against the screen and held there. This possibility should be avoided particularly where bypass location and design does not fully meet the requirements noted in Section 6.4 of this chapter. Fish that are delayed for long periods of time because they are unable to find the bypass are also likely to eventually drift against the screen and remain there until they die if velocities through the screen are too high.

In speaking of the open space through a screen the *net* open area is intended, of course. This means that the area of the frame and any cross supports as well as the area of the wires themselves must be subtracted from the gross area normal to the direction of the flow. In this connection, galvanizing the screen also increases the area of the wires and reduces the net area. It is usually possible, as noted above, to design the screen and frame support so that a clear opening of at least 50% of the gross area remains, however.

The head loss through the screen, which increases with a decrease in the net open area, can be an important consideration to the owner of the intake. For irrigation ditches operating by gravity flow, a few inches of head loss can seriously reduce the quantity of water available for use by the farmer or rancher. For other intakes where there is ample gravity head or pump capacity, a small loss of head may not be serious. The head loss at the lowest approach velocity specified, 0.4 ft/sec, is almost negligible. For drum screens of the mesh sizes specified for spring and coho salmon fry, the total head loss with a 50% clear opening at an approach velocity of 0.4 ft/sec is less than 0.02 ft. For a single vertical screen it is less than 0.01 ft. The head loss through a single screen increases rapidly with increased velocity as shown in Figure 6.6, however.

Figure 6.6 also illustrates the difference in head loss through a screen with a change in percentage opening. A screen with mesh openings as specified for spring and coho salmon fry, but with a total clear opening of only 30%, will have a head loss five times that of the 50% open screen at an approach velocity of 1.0 ft/sec.

Screens of larger mesh opening and larger percentage opening will have less head loss for comparable approach velocities than those shown, while screens of smaller mesh opening, such as the one specified for the smallest salmon fry, will have only a very slightly larger head loss, provided that the percentage opening is maintained at a minimum of 50%.

For practical purposes, the velocity through the screen and the resulting head loss depend more on the amount of debris collected on the screen at any given time. This is where the designer must exercise judgment and draw on available experience in any particular location. For vertical screens with no provision for automatic cleaning, the head loss due to the clean screen is the minimum to be expected, and under normal operating conditions it will be much higher on the average even if the screen is frequently cleaned. For screens with provision for automatic cleaning, such as the drum screen or endless-belt traveling screens described later in this chapter, the approach velocity can be used directly for determining screen area. But where cleaning is performed manually and is likely to be infrequent, such as only once or

Approach velocity in cm/sec.

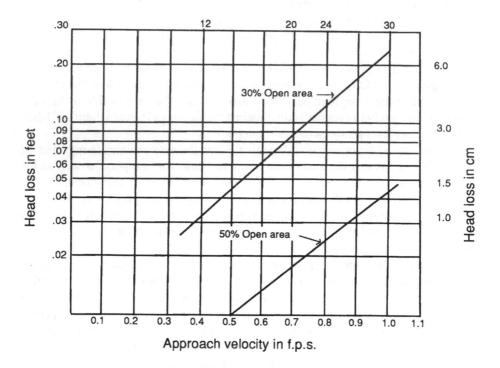

Figure 6.6 **Graph showing head loss through single clean wire screens with square openings about 0.12 in. (0.3 cm) on a side at various approach velocities for screens with 30 and 50% net open area.**

twice daily or at longer intervals, additional screen area is advisable. This allows a margin of safety on screens for small pump intakes and in general for intakes in facilities where it is not economical for the owner to employ an attendant to clean the screen regularly. A 50% clear opening is required on all screens, of course, whether manually or automatically cleaned.

Screens other than wire mesh have also been used for preventing entry of fish. Thin steel plates with perforations of various sizes and shapes are available commercially. Some of the shapes of perforation available are round, square, hexagonal, and slotted. These are available in various plate thicknesses and in materials other than steel, such as copper, brass, zinc, etc. In general, perforated plates will have a lower coefficient of discharge than a wire screen with equivalent size of mesh opening, and will thus have a greater head loss. They are not generally used for fish screens except in special cases, such as where a mechanical wiper is attached to keep the screen free from debris.

A newer development is the wedge wire screen shown in Figure 6.7. Manufactured by the Johnson Company of St. Paul, MN, it has flow-through characteristics improved over the wire mesh screen. It is also largely self-cleaning when used as a sloped installation with approach velocities **over 3 ft/sec**, and it diverts fish more

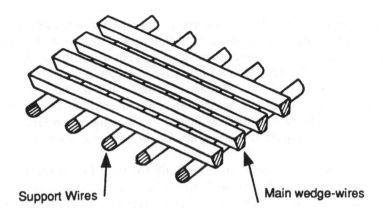

Support Wires   **Main wedge-wires**

Figure 6.7 Section of stainless steel wedge wire screen.

readily without impingement of the fish on the screen. While information about head losses is not presently available, the screen shows promise as will be seen by the variety of installations using it, described later in this chapter.

6.6 THE BYPASS

Bypasses have been incorporated in fish screen installations over the years that screens have been in use. Improvements in bypass design have been developed even more slowly than in screen design, however. It is suspected that the reason for this has been that the screen has usually provided a definite barrier to any further downstream migration, and the provision of any kind of bypass is considered to be ample to permit the fish to return to their normal migration route down the river or stream. However, recent research conducted on other types of diverters or deflectors has given rise to a reconsideration of bypass requirements.

Chief among these more recent experiments were those with louver diverters. These experiments showed a need for better bypass design. Briefly, it was found that fish diverted during the experiments that did not readily find the bypass would return upstream a short distance and then repeat their approach to the diverter. After a number of approaches, more and more of the fish would penetrate the diverter. It appeared logical that if the fish, after first being diverted, were carried into a carefully designed bypass that prevented them from swimming back out, losses through the diverter would be considerably reduced. This subsequently proved true, and the efficiency of the diverters was greatly increased by improvements in the design of the bypasses.

It is now believed that similar improvements in design of bypasses at screens that form a *complete* physical barrier are just as beneficial. Fish that are not immediately carried into a bypass after being deflected by a screen can try to penetrate the screen again, but cannot pass through it. However, they can become fatigued after many successive attempts and will eventually die, or at least become vulnerable to predators as a result. The improvement of bypass design for screens can therefore result in decreased mortality rates for fish that must be deflected by the installation.

Ideally, the design of a bypass is keyed to the behavior of the fish it is desired to save. However, there are pronounced differences in behavior of downstream migrant juveniles of different species, and it is often physically impossible to reproduce conditions ideally suited to each of the several species that might be present at one screen. An example of this difference in behavior is the schooling of sockeye salmon smolts as compared to the nonschooling of coho and chinook salmon smolts. For the sockeye smolts a wide, shallow bypass might be the best design in order to capture the fish in complete schools, while for the other species a narrow bypass could be used.

There are also differences in the swimming ability of fish of varying sizes. A bypass that might successfully capture Pacific salmon fry might be unsatisfactory for smolts of the same species, particularly for two-year-old smolts, which are much larger than fry. These fish can swim much faster and might easily escape from the bypass if velocities are not high enough.

Some of the principles to be kept in mind in designing a screen bypass are as follows:

1. **Location** — The bypass entrance should be located as near as possible to the area where the fish are stopped by the screen. Thus it is usually located at or near one end of the screen if the screen is short, as in Figure 6.3, or at each end if a longer screen is used. If the screen is extremely long, it might be advisable to place bypass entrances at intervals across the face. This has been accomplished by incorporating the entrances in the support piers in those cases where a number of screen panels have been placed across a diversion canal. These entrances are connected to a pipe buried in the floor of the installation, which carries the bypass water back to the river. Bypass entrances have usually been placed near the surface, because most screens are installed in fairly shallow water, and a surface or near-surface entrance is reasonably accessible to the blocked fish. Where the screen is short and in deep water, as in the case of a mechanical, traveling screen, it might be advisable to have additional submerged bypass entrances. This is particularly desirable if the water is drawn through the screen at some depth and there is a nonuniform approach velocity, with much lower velocities at the surface.

2. **Velocity** — Simple bypasses for small irrigation intakes usually consist of a weir constructed across the widened entrance to the bypass as shown in Figure 6.3. A notch or orifice in the weir permits the desired flow to pass over or through it and then fall into a sump or well behind the weir, to be carried back to the river through a pipe or flume. If the head between the water surface in the sump and the water surface in front of the screen can be maintained at 18 in. or more, it is unlikely that even the largest smolts can return to the diversion canal after passing over the weir. The attractive velocities to the bypass entrance are not great for normal installations of this type, but the economically justifiable bypass flow, in cases where this simple installation is used, is not likely to be great enough in any case to provide more attractive velocities by changes in design. For practical

reasons, this type of bypass is good, because its simplicity of operation enables the owner, who is usually a farmer or rancher, to maintain the required bypass flow by simple adjustments without wasting water needed for his or her crops.

Ideally, however, a bypass should have a volume of flow as large as possible in order that its cross-sectional area at the entrance can be large and still have satisfactory velocities in a downstream direction. Brett and Alderdice (1953) recommend among other things that for schooling fish the bypass should be wide, that the flow through it should be "stream-lined," and that it should be uniformly accelerating at a rate not more than 0.1/ft of length. Ruggles and Ryan (1964), from their experiments with louvers, recommend that for schooling fish the bypass entrance be a minimum of 18 in. (45 cm) wide, and the velocity at the mouth be at least equal to the approach velocity to the louver and preferably 40% greater. Such ideal conditions may not always be attainable.

The gradual acceleration recommended by Brett and Alderdice is intended to eliminate sudden changes in the environment, which might scare the fish away from the bypass. Uniform velocities over the cross-sectional area of the bypass are also important, hence the recommendation for "streamlined" flow. Areas of noticeable turbulence caused by projections into the bypass, and differences in velocity from top to bottom or from side to side of the bypass channel are not desirable. Besides disturbing the fish, they result in localized areas of low velocity where the fish can rest before swimming back upstream.

3. **Quantity of flow** — As noted previously, the larger the bypass flow, the more attractive it is to the fish and therefore the more successful it is likely to be. However, if the entrance is located close to the front of the screen, the bypass flow need not be excessive. For larger diversions it is unlikely that the bypass flow needs to be more than 1% of the total. For smaller diversions, the proportion would be greater, since extremely small flows in the bypass would not provide enough water for the fish to swim in. For these smaller installations, it is doubtful if it would be desirable to provide anything smaller than an 8- to 10-in.-diameter pipe to carry the fish back to the river. While this pipe need not necessarily run full, it should carry sufficient water to provide adequate swimming depth. Since gradient and pipe materials will differ, it is the responsibility of the designer to adjust these variables to the quantity of flow from the bypass to give the desired depth of flow in the pipe.

6.7 SCREENS FOR IRRIGATION DITCHES — THE REVOLVING DRUM SCREEN

The revolving drum screen was developed by the Oregon Game Commission in 1921 for use in irrigation ditches in that state. Since that time it has been used in hundreds of similar installations on the Pacific Coast of North America with only minor improvements. Figure 6.8 shows an installation in a small irrigation ditch. The screen drum is 6 ft long and 2 ft 6 in. in diameter. When in operation and carrying water to a depth of about 2 ft, this installation can effectively screen 6 cfs of water having an approach velocity of 0.5 ft/sec, or 12 cfs with an approach velocity of 1.0 ft/sec. The water velocity downstream from the screen is sufficient to drive the paddle wheel, which in turn rotates the drum screen at a much slower rate in the opposite direction by a chain and system of reduction gears. The screen is protected from heavy trash by a simple timber trash rack placed just upstream, and a bypass is provided with its entrance adjacent to one end of the screen. The entire assembly is housed in a simple, open concrete box with 6-in. walls and floor lightly reinforced. Some care is required in the construction, adjustment, and operation of this type of screen. The drum must be fitted carefully in the box to eliminate any spaces around the edges larger than the openings in the screen mesh. The screen must be carefully fastened to the drum so as to avoid bulges or depressions that would create a space as the drum turns. If the screen is pulled too tight on the drum, it will bow between supports, leaving a space through which fish might escape. Rubber seals must be provided along the base and sides to take care of minor irregularities in the screen surface. These precautions are all necessary because fish seem able to very quickly find any gap that is accessible and large enough to pass through.

A weakness of this type of screen is the fact that it cannot be submerged to a depth approaching its full diameter without providing an "escalator" for fish to travel over the screen with the debris. In normal operation with water at a depth of two thirds or three quarters of its diameter, the screen carries debris over as it turns, and this is washed off as the screen is reversed on the downstream side. It will be seen, however, that as the depth on the drum increases, the curved surface of the screen approaches the horizontal at the point it leaves the water surface. Fish seem to have a tendency to lie against the screen in the low velocities at the surface, and are readily carried over the screen with the debris if the depth is greater than the desirable operating range.

This type of screen is simple and economical to construct, operate, and maintain, and for this reason it has been very popular in the Pacific drainage where considerable areas are under irrigation. A screen that can operate at approach velocities of about 1 ft/sec costs about $2000/cfs to install for installations of 10 cfs or more. For lower velocities the installation costs per unit of water screened are proportionately higher. Maintenance costs are about 2% of the capital costs per year. The chief maintenance costs result from the need to replace the seals and repaint the metal parts to prevent rust. A permanent timber frame is usually provided over the screen drum to assist in lifting the drum up for inspection and maintenance. In addition, a sandtrap, consisting of a depression in the floor ahead of the screen, is provided to keep the installation free of sand and silt, which will not only interfere with its day-to-day operation but greatly add to wear of the moving parts.

While the drum of the screen shown is driven by chain and sprocket, it is also possible to drive these screens by other linkages such as a drive shaft and bevel gears. If electricity is available, a small motor can be substituted for the paddle wheel. For larger installations, several drums can be operated side by side across the canal, with a bypass opening incorporated between adjacent drums.

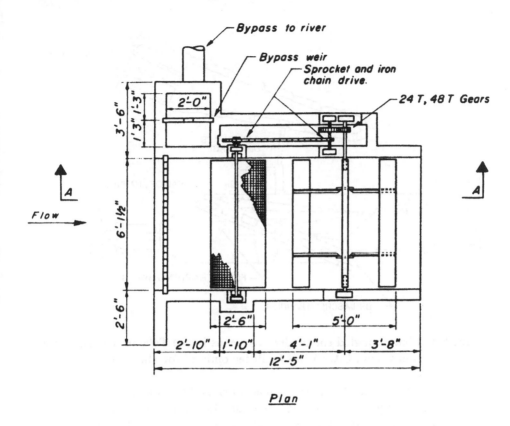

Plan

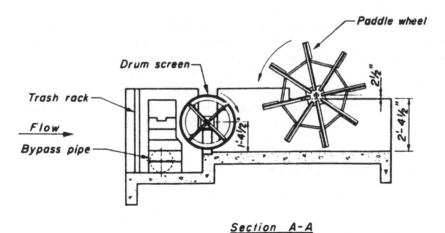

Section A-A

Figure 6.8 Revolving drum screen driven by paddle wheel, as used on small irrigation diversions on the Pacific Coast of North America.

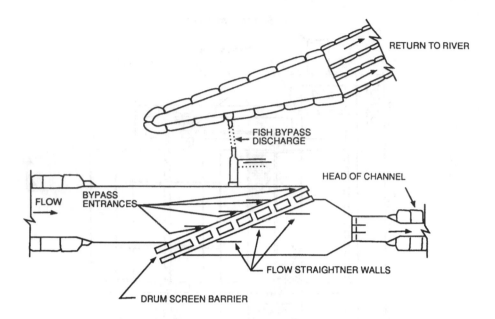

**Figure 6.9   Plan of angled rotary drum screens in the Tehama-Colusa Irriga-
tion Canal, CA. (Adult facilities omitted for clarity.)**

This type of screen, while most popular for small irrigation diversions, has also
been used for fairly large diversions at hydroelectric power developments. An
example is the installation on the White River by Puget Sound Power and Light,
which screens 2000 cfs (57 m³/sec).

The screens described so far have been installed perpendicular to the canal or
intake channel. More recent experience suggests that angling the screen across the
canal and providing better bypass velocities improves the effectiveness considerably.
Figure 6.9 shows one such installation on the Tehama-Colusa Canal in California.
The multiple bypasses will be noted. It is installed together with facilities for adult
fish, which are omitted from the diagram for clarity. Several large new installations
of this type are planned for the future on the Pacific Coast.

6.8 INCLINED PLANE SCREENS

The first inclined plane screen may have been the one described by Wales et al.
(1950). It was installed in an irrigation ditch and consisted of a smooth perforated
plate sloped at an angle of 33½° to the horizontal. The original had a paddle-wheel-
driven arm with a large wiper blade attached, which scraped debris over the screen
as it completed its periodic up-and-down movement. This was not very satisfactory,
as it carried fish both dead and alive with the trash over the screen. It was, however,
the forerunner of more successful installations.

For some intakes and specialized use, Kupka (1966) developed a more elaborate version
with the screen at an even smaller angle to the horizontal and with brushes on an endless belt,
brushing the screen from bottom to top. The fish and debris were passed over the screen, and

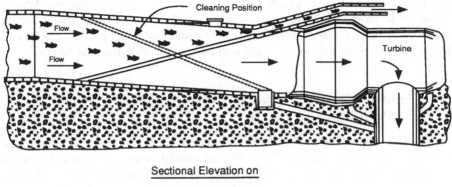

Sectional Elevation on

Centre-line of Penstock

**Figure 6.10   Pressure wedge-wire screen in turbine penstock of T.W. Sullivan
Plant at Willamette Falls, OR, screening 425 cfs (12 m³/s). A
sectional elevation on the center line of the penstock. (From Eicher,
G.J., 1982; patented by G.J. Eicher; used with permission.)**

the debris was separated from the fish so that it fell in a special trough, while the fish fell in
another trough leading to a bypass. This is reported to have worked satisfactorily under the
conditions occurring at Robertson Creek in British Columbia. Since the angle of the screen
with the horizontal was so small, fish were not impinged on it and were easily carried over
the screen to the bypass.

Another installation, described by Finnegan (1977), is completely horizontal and
submerged, so that the problem of debris is almost eliminated. In spite of this, it has
a rotating paddle wheel driving bristles to keep it clean.

The most important factor affecting development of the inclined plane screen,
however, was the introduction of the wedge wire screen. Its superior hydraulic
properties have contributed greatly to the further development of this screen as well
as making possible other applications, which will be described later.

Eicher (1982) describes an installation in the T.W. Sullivan Plant at Willamette
Falls, OR, which includes a sloping screen installed in a penstock as shown in Figure
6.10. The screen, composed of wedge wire bars 0.08 in. (2 mm), spaced 0.08 in.
(2 mm) apart, is placed at an angle of 19° to the flow in the penstock. Average
velocity in the penstock is 5 ft/sec (1.5 m/sec), and velocities normal to the screen
surface are 1.5 ft/sec (0.46 m/sec). The high velocity carries young salmonids into the
bypass, and debris collection on the screen is almost eliminated. The screen can be
reversed for cleaning, as shown in the diagram, but this is seldom necessary. Survival
of young salmonids diverted through the bypass is reported to be 90%. Improvements
to be made in roughness through the penstock and bypass to reduce descaling are
expected to improve survival.

This installation has not gained general acceptance by the responsible regulating
agencies, however, since it deviates from their velocity criteria of 0.4 to 0.5 ft/sec at
the face of the screen.

Further experimentation was done with inclined screens of perforated aluminum
plate, and experiments were conducted in the late 1970s with perforated plate and

mesh plate using other species such as alewife, striped bass, white perch, and Atlantic tomcod, with promising results. This was extended to wedge wire screens for the protection of fish eggs and larval stages, with equally promising results. It appears likely that further research in this field will be conducted in the near future.

6.9 SCREENS FOR INDUSTRIAL INTAKES — FIXED, PERPENDICULAR TO FLOW

Fixed or stationary screens have been used successfully for industrial and domestic water-supply intakes. They require careful design, however, to ensure satisfactory service, and must be installed with a full realization of the cleaning problem, which in most cases involves a large amount of manual labor. The general design criteria outlined previously apply to such screens, except that a large safety factor in screen area is desirable.

For example, if a screen is required to prevent entry of downstream migrant salmon fry or other fish of a similar length of about 1 in. (2.5 cm), it might be decided that screen made of wire 0.028 in. in diameter with eight meshes to the inch will be satisfactory. The velocity of approach can be 0.4 to 0.5 ft/sec, and the velocity through the screen would be about double this since this mesh gives about a 50% clear opening. However, conditions at the installation may be such that an attendant can only be provided to check and maintain it once a day at the most. In many installations the intake and screen are a considerable distance (perhaps several miles) from the point where the water is used, and the volume of water involved is not sufficient to justify employing a worker full time to maintain the screens. In such cases it could be anticipated that the screen will become partially blocked by debris, algae, etc., between cleaning periods. Assuming the water drawn into the intake remains constant, the mean approach velocity will remain the same with a partially blocked screen, but the velocity through the screen will increase greatly. In addition, there are likely to be localized areas of high approach velocity and of high velocity through the screen where there are clean patches. In such cases it has been found desirable to specify sufficient screen area to provide for approach velocities as low as 0.10 ft/sec. This permits the screen to have as much as three quarters of its surface area blocked without creating localized areas where the approach velocities exceed 0.4 ft/sec.

For most small installations such as pumped supplies for small irrigation and domestic water intakes, a simple box with the required screen area is not unduly expensive and is fairly easily maintained. For large installations such as intakes for industrial processes and cooling water, a continuous water supply is sufficiently important to justify regular inspection and maintenance.

It is also usual to specify that a coarser protective screening be placed outside the fine screen, in cases where a trash rack with close bar spacing is not used. This protects the fine screen from being damaged by large pieces of debris. A screen with wire diameter of around 0.072 in., and with three meshes to the inch is often adequate for this, but other combinations of wire size and mesh might be used, depending on the desired strength and span of the panels. Care must be taken, of course, that the percentage of open area through this coarse screen is somewhat greater than that

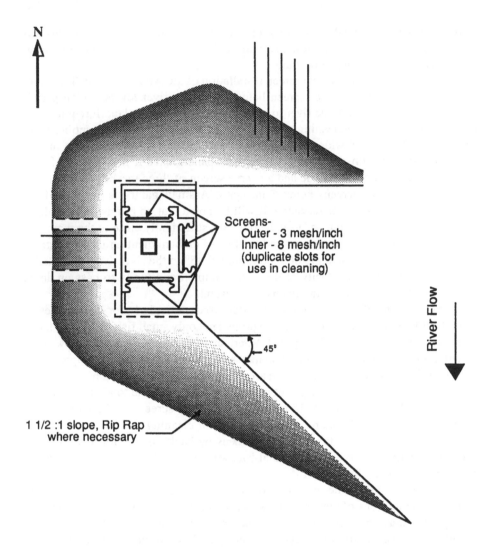

Figure 6.11 **Plan of a fixed screen intake for cooling water from a river. Screen panels are removable for cleaning.**

through the fine screen, because a high velocity through the coarse screen could result in fish becoming trapped between the two screens.

Provision for the cleaning operation of such screen is important. The screens for larger installations are usually made up in panels. These can be constructed of a frame of steel angles or channels, with the fine screen stretched across the inside and the coarse screen across the outside. These panels can then be slipped into guide slots in the box housing the screen. Care should be taken to ensure that the panels fit snugly enough in the guides so that spaces larger than the clear opening in the fine screen mesh do not occur, as these spaces could result in loss of fish around the screen. In order to protect the fish at all times, it is desirable to have a double set of slots, side by side, for the screen panels. A spare screen panel is then provided, which is slipped into the spare slots while the panel to be cleaned is removed for servicing. In this way, losses are reduced to just the few fish that might become trapped between the two slots during the foregoing procedure.

Figure 6.11 shows a typical screen installation of the type described. This intake provides an area of 100 ft² of screen below minimum river levels, which permits withdrawal of 10 cfs from the river to cool a diesel generating plant. It requires only occasional removal and cleaning of the screen panels. The screen assemblies themselves have frames of wide-flange beams, with three removable panels in each assembly framed with angles and plates. The panels have a fine screen of eight meshes to the inch on the inside and a coarse screen of three meshes to the inch on the outside, the two screens being 5 in. apart. A small gantry is used to lift the assembly from the slots for servicing. It will be noted that no provision is made for a bypass at this installation, because the intake is sufficiently close to the main flow of the river that it is not necessary, and the line of embankment on the downstream side of the screen is at an angle of approximately 45° to the river. This permits fish that are attracted to the velocity of the screen ample opportunity to pass on downstream without having to swim against the current.

Fixed screens of various forms have been used particularly for lake intakes for a number of years. Here it is necessary to protect not only small fish in the 1 in. length category, but even smaller life stages of American shad, rainbow smelt, yellow perch, alewife, shiner and/or other species that may be present in the lake. The approach has been to make the screen openings as small as necessary (down to 0.02 in., or 0.5 mm) and to reduce the velocity through the screen (by increasing the total area of screen) to a range such as 0.5 ft/sec, or 15 cm/sec. Wedge wire screens of these dimensions have proved successful, and require a minimum of cleaning.

The usual practice is to clean such screens by backflushing for a short period (say, 10 s) daily. In some cases thermal backwashing has been tried, but it has not been entirely successful.

It would seem that fine-mesh low-velocity screens have to be adapted in design to the local conditions of algal growth, species present, size of species, temperature, quality of water, etc. Screens of this type are more sensitive to passive growth conditions in the environment than are screens of wider mesh opening in river intakes.

The costs of fixed screens are comparatively low, but still vary a great deal according to the area of screen, velocity of flow, and maintenance expected. There is therefore, little point in rationalizing their cost on a "unit of water screened" basis. As pointed out, they have a high maintenance cost, which should be taken into consideration when deciding on this type of installation.

6.10 SCREENS FOR INDUSTRIAL INTAKES — MECHANICAL OR SELF-CLEANING

Because of the labor costs involved in manually cleaning fixed screens, it was inevitable that industry would look to some type of mechanical, self-cleaning screen to lower the operating costs. Many types have been developed for this purpose,

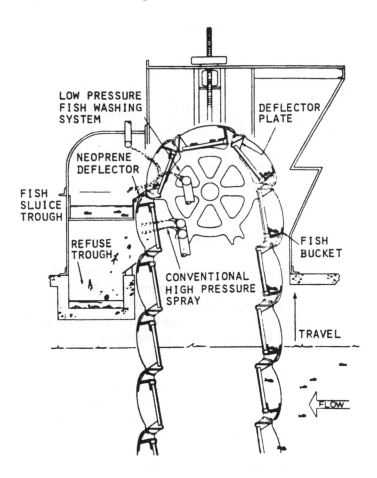

Figure 6.12 A traveling screen modified for protection of fish.

including bar screens with automatic rakes, endless-belt traveling screens which utilize jets of water for washing the screen, and various types of drum screens also utilizing water jets for cleaning. While these screens have been used for many industrial purposes, particularly in the domestic water-supply, sewerage, and paper-processing fields, only the endless-belt or traveling water screens have been successfully adapted for use as fish screens up to the present.

Figure 6.12 shows a section through a traveling water screen, which has been modified for protection of young fish. The screen itself is a standard flow-through water screen, and it may have mesh of the sizes outlined previously for small salmon, or it can have mesh sizes considerably smaller down to 0.02 in. (0.5 mm), to exclude the young of alewife, walleye, striped bass, etc., down to the larval stage. It has been found in tests of screens of this type that if they are continuously operated, have low-pressure sprays, and fish-lifting buckets fastened to the screen panels, they can protect relatively fragile fish satisfactorily.

The lips or buckets for holding the fish that become impinged on the screen as it lifts out of the water will be noted. They are made watertight and give the fish

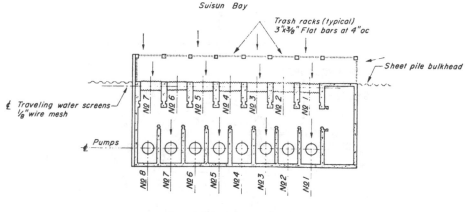

Figure 6.13   Plan of cooling water intake for the Pittsburgh Steam Plant of the Pacific Gas and Electric Company in California. The traveling screens, which are set flush with the outside wall of the pumphouse, pass up to 900 cfs. Fish can pass freely through the trash rack that projects out from the structure. (After Kerr, 1953.)

moisture as they travel over the screen to the jets. The low-pressure jet then washes them into the fish sluice trough. The high-pressure jet then cleans the screen before it goes down to repeat the cycle. All these features normally can be retrofitted on a standard traveling screen and improve its effect on fish. Impingement used to be a chief source of overall mortality at this type of screen, and it can be greatly improved by these additions.

Overall survival of even the most delicate of young fish has been reported in the 95% range in tests at various locations.

A bypass is necessary in this type of installation when the screen is located some distance from the main body of water supplying it. The design of these bypasses should follow as nearly as possible the principles set out previously. However, because of the practical limitations governing bypass efficiency, it is desirable and often possible to avoid the need for a bypass by placing the screens near the entrance to the intake. Figure 6.13 shows the plan of the cooling water intake for the Pittsburgh Steam Plant of the Pacific Gas and Electric Co. in California. According to Kerr (1953), this intake screens up to 900 cfs of water by the use of mechanical traveling screens located flush with the face of the intake structure. The trash rack projects out from this face, allowing the waters of Suisun Bay to circulate freely past the screen face. If approach velocities are kept low, this is almost an ideal layout for a screen installation, eliminating the need for a bypass completely.

Center-flow screens have also been utilized for industrial intakes, mainly in Europe, and to a lesser extent in the U.S. It has been found that these too can be modified to reduce impingement mortality to fish. The method is essentially the same, that is, the addition of fish buckets to the screen panels and the use of a gentle spray wash at the top of the screen.

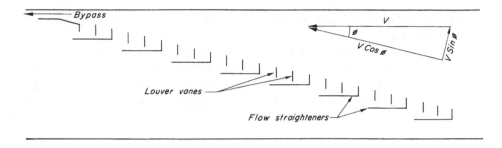

V = Approach velocity of water in flume
ϕ = Angle of line of louvers
$V \sin \phi$ = Swimming velocity of fish
$V \cos \phi$ = Resultant velocity of body of fish

**Figure 6.14   Plan of louver array mounted in flume, showing theoretical veloci-
ties of flow and resultant speeds of fish movement.**

6.11 LOUVER DIVERTERS

The louver diverter, which was newly developed and showed great promise at
the time of writing the original 1961 text of this book, has unfortunately not met with
the success one would have expected over the intervening period.

It has been extensively tested with many species of fish, including salmonids,
striped bass, white catfish, threadfin shad, northern anchovy, queenfish, white croaker,
walleye, and shiner perch. The outcome of this testing has varied greatly, with
efficiency of diversion ranging from 40 to over 90%. Efficiencies in the 90% range
were obtained only for salmon migrants, while lower efficiencies were obtained for
migrants such as striped bass and some other species.

Thus it is apparent that unless further development changes this, louvers will be
used only in cases where near-100% guiding efficiency, as is possible with screens,
is not needed. One such case is at a power development intake, where it is known that
only a small percentage of fish are killed in passing through the turbines. If a louver
is used in such a case, it can successfully reduce these losses to a fraction of the
original. For example, if the turbine loss at a dam is 10%, and louvers are used ahead
of the turbines with a guiding efficiency of 90%, then the turbine loss can be reduced
to only 1%.

For the engineer or biologist who is interested in considering this option, the
following description is offered. The louver array consists of a series of bars mounted
in a flume vertically, with the flat side of the bars at right angles to the flume walls
so that they are normal to the flow in the flume, as shown in Figure 6.14. They are
mounted in a row, which is usually set at a very shallow angle to the flow, such as
15°. This angle is shown as ϕ in the diagram. The velocity components are also
shown in the diagram. V is the main approach velocity, $V \sin \phi$ is the velocity normal
to the louver and is kept at or below the swimming speed of the fish. The louver
operates on the assumption that fish approach it tail first, maintaining themselves thus
by swimming at a certain speed that is less than V but being carried downstream
passively by the current. When they sense the effect of the louver, they redirect their

velocity in the direction $V \sin \phi$, and thus are carried in the direction $V \cos \phi$ until they reach the bypass.

In the tests conducted to date, the louver bar spacing has been varied from 1 in. up to 4 in. or more, the approach velocity has been varied, the bypass velocity has been varied, and the species have been varied, producing a welter of data, very little of which has shown any consistency except the very general ones I will outline as follows:

1. For salmonids, varying the approach velocity has very little effect on guiding efficiency. In most tests, this velocity was from 1.5 to 4 ft/sec. This is an important consideration for the designer of the flume, whether it is a power intake flume or for some other purpose.
2. For other species, varying the approach velocity did have an effect on guiding efficiency. Fish such as striped bass and white catfish, which migrated at a length much less than salmonids, and whose swimming ability at that size was much lower, showed a definite inverse relationship between efficiency and approach velocity. Guidance efficiency was much improved when approach velocity was less than 2.5 ft/sec.
3. Bar spacing seemed to have little effect on guiding efficiency of salmon smolts. For these, bar spacing up to 6 in. gave efficiency in the same range as less spacing. However, bar spacing became important for salmon fry and for the young of the other species studied. For salmon fry, bar spacing of not more than 2 in. is recommended as a result of one set of experiments (Ruggles and Ryan, 1964). Bar spacing for other species has been mostly tested at 1 in.

Besides these general recommendations, the considerable amount of louver testing to date has resulted in a much better understanding of bypass requirements, which can be usefully applied in design of all screening and diverting installations.

The first important requirement in bypass design is that the width must be consistent with the behavior of the fish. If the fish are schooling, the bypass must be wider, since it takes only a few fish turning back when the walls get too close, to get all the fish to turn back and swim upstream.

A minimum of 18 in. (45 cm) has been recommended in such cases. For nonschooling fish, 6 in. (15 cm) can be satisfactory.

The second important requirement in bypass design concerns the velocity. It was found that the bypass velocity must increase over the approach velocity in the flume, or the guiding efficiency decreases. Various figures are offered for this increase, but in general, an increase of 140% of the approach velocity in the flume is considered a safe requirement.

The costs of louver installations are generally lower than the cost of screens, because the installation is comparatively simple to build. The costs of louver maintenance are usually much lower than for screens, because of the relatively wide bar spacing, which is less susceptible to trash accumulation. Nevertheless, they do require periodic maintenance, and the costs of operation lie somewhere between those of screens and of normal widely spaced trash racks.

6.12 PARTIAL SOLUTIONS TO PROTECTING DOWNSTREAM MIGRANTS AT HYDROELECTRIC INTAKES

Many hydroelectric dams were built 40 to 50 years ago, before the mortality of young fish passing through their turbines was measured or even considered. Attempts are now being made to retrofit facilities to reduce these mortalities where they have come to be considered critical to the continuing existence of the migratory fish.

At some installations on the East Coast of North America and in Europe, a bypass has been provided to form a route for downstream migration, with varying degrees of success. The bypass usually consists of an outlet channel at the surface, leading to a pipe or gently sloping channel to carry the fish down to the tailrace without injury. It has been most successful where the layout of the dam provides a favorable location for the bypass entrance, that is, where the downstream migrant fish will readily find it, and where they do not have to sound, but are able to pass into a surface entrance rather than to the turbine intakes, which are usually submerged.

Other intakes have been partially screened or screened with an apparatus that is only partially effective, such as a louver. This has been the case on both the East and West Coasts of North America. At Malay Falls, in Nova Scotia, Canada, Semple (1979) describes the testing of an installation of a louver screen and a partial vertical screen on two dams to determine if the partial screen is effective at partially offsetting the turbine mortality of 10%. His conclusion was that the bypass flume alone reduced this mortality to approximately 5% by diverting approximately 50% of the descending Atlantic salmon smolts. This could possibly be improved by adding a rigid frame screen deflector in front of the turbine intake, leading to the bypass intake. At another installation, Semple (1977) describes the sampling by video television and sonar of a run of adult alewives through a bypass around a dam at Tusket Falls. The alewives had spawned above the dam and were returning downstream. They numbered 144,000. The bypass also passed a run of salmon smolts.

On the West Coast of North America, at the many dams on the Columbia River, partial screens have been tested and found satisfactory for diverting many of the migrant salmonids from the entrance of the turbines to safe passages through the gatewells and thence by pipe or truck transport safely downstream. These installations have been mostly submerged traveling screens, although in some cases stationary wedge screens have been tested.

Installations have been made at McNary Dam, Bonneville Dam, John Day Dam, and Rocky Reach Dam on the Columbia River, and at Little Goose and Lower Granite Dams on the Snake River. These have all been basically a submerged traveling screen as shown in Figure 6.15. According to the report of Stone & Webster Engineering Corp. (1986), quoted earlier in this chapter, at a typical installation at McNary Dam, the mesh size of the submerged traveling screen (STS) is 0.25 in. (0.64 cm). The fish, which are concentrated mainly in the upper levels of the water, are deflected upward into the gatewells. A barrier screen of the same mesh size guides the fish to submerged orifices leading to a collection channel, from where they are trucked or barged below the dam.

Mortality in this deflection system averages 1%. Fish-guidance efficiency has been estimated as 74% for spring chinook and 38% for underyearling chinook.

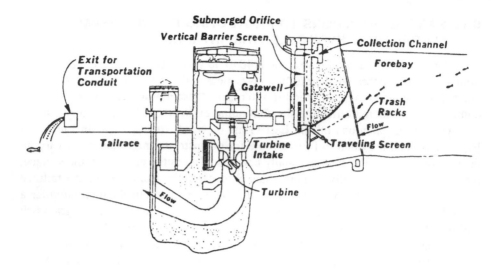

Figure 6.15 Submerged traveling screen as installed on the Columbia River dams.

Similar installations have been made or are planned at the other dams noted on the Columbia and Snake Rivers. These have had numerous problems in maintenance and operation, since they are so complex. But the problems have not been insurmountable, and the screens have been in continuous service from April to October or whenever the local migrations occur. The cost has been fairly high, ranging from $7 million to $11 million per dam.

6.13 LITERATURE CITED

Bainbridge, R., 1960. Speed and stamina in three fish, *J. Exp. Biol.*, 37(1), pp. 129-153.

Bell, M.C., 1981. Updated Compendium on the Success of Passage of Small Fish through Turbines, U.S. Army Corps of Engineers, North Pac. Div., Portland, OR. 294 pp. plus tables.

Bell, M.C. and A.C. DeLacey, 1972. A Compendium on the Survival of Fish Passing through Spillways and Conduits, U.S. Army Corps of Engineers, North Pac. Div., Portland, OR. 121 pp.

Brett, J.R. and D.F. Alderdice, 1953. Research on Guiding Young Salmon at Two British Columbia Field Stations, Bull. No. 117, Fish. Res. Board Can. 75 pp.

Brett, J.R., M. Hollands, and D.F. Alderdice, 1958. The effect of temperature on the cruising speed of young sockeye and coho salmon, *J. Fish. Res. Board Can.*, 15(4), pp. 587–605.

Canadian Electrical Association, 1984. Fish Diversionary Techniques for Hydroelectric Turbine Intakes, Montreal Eng. Co., Rep. No. 149G399. pp.

Eicher, G.J., 1982. A passive fish screen for hydroelectric turbines, ASCE Hydr. Div. Mtg., Jackson, MS, August, 1982.

Finnegan, R.J., 1977. Development of Self-Cleaning Fish Screen, Can. Fish. Oceans, Vancouver, B.C. 2 pp. plus 5 fig.

Kerr, J.E., 1953. Studies on Fish Preservation at the Contra Costa Steam Plant of the Pacific Gas and Electric Co., Calif. Fish & Game, Fish. Bull. No. 92. 66 pp.

Kupka, K.H., 1966. A downstream migrant diversion screen, *Can. Fish Cult.,* 37, pp. 27–34.

Larinier, M., 1987. Mise au point d'un protocole experimentale pour l'evaluation des dommages subis par les juveniles lors de leve transit a traversu des turbines, CEMAGREF Convention DPN No. 85/8. 20 pp. plus tables.

Ruggles, C.P., 1980. A Review of the Downstream Migration of Atlantic Salmon, Can. Tech. Rep. Fish Aquatic Sci. No. 952. 37 pp.

Ruggles, C.P. and P. Ryan, 1964. An Investigation of Louvers as a Method of Guiding Juvenile Pacific Salmon, Dept. Fish. Can., Vancouver, B.C. 79 pp.

Semple, J.R., 1977. Video Television and Sonar Sampling Techniques in the Study of Adult Alewives at a Hydroelectric Dam Bypass, Tech. Rep. No. MAR/T-77-1, Dept. Fish. Oceans, Halifax, N.S.

Semple, J.R., 1979. Downstream Migration Facilities and Turbine Mortality Evaluation, Atlantic Salmon Smolts at Malay Falls, N.S. Dept. Fish. Marine, Manu. Rep. No. 1541, Halifax, N.S.

Stone & Webster Engineering Corp. 1986. Assessment of Downstream Migrant Fish Protection Technologies for Hydroelectric Application, Final Rep. to Electric Power Research Institute, EPRI AP-4711, Project 2694-1, Palo Alto, CA.

Wales, J.H., E.W. Murphy, and J. Handley, 1950. Perforated plate fish screens, *Calif. Fish Game,* 36(4), pp. 392–403.

7

FISH PASSAGE
THROUGH ROAD CULVERTS

7.1 THE PROBLEM

While this text is intended to cover the more formal types of fish facility, there are many other migratory fish problems that the fishery engineer and biologist face in their day-to-day work. We include here a detailed example of one of these and explain how it is being handled in many different locations and with many different species of fish.

One problem that occurs increasingly frequently is the blockage of fish passage at road culverts. Road and highway construction has increased rapidly in most countries, and shows no signs of slowing down. Road design and construction involves the design of bridges and culverts over every stream crossing, and while bridges are preferable for providing fish passage, the development of new types of piping for culverts has encouraged engineers to give preference to culverts rather than bridges, because of the savings in cost.

7.2 BIOLOGICAL ASPECTS

At the same time, fishery biologists have become increasingly aware of the need for ensuring that these culverts are passable to fish migrating particularly in an upstream direction. Surveys have shown in many cases a difference in fish populations in the streams above and below existing culverts, leading to the conclusion that free passage is not possible through them.

In order to demonstrate the need for fish passage, the biologist must determine what species of fish are unable to ascend an existing culvert, or in the case of new culverts, what species are likely to require passage because of migrations that may occur at certain times during the life of the species. The following U.S. Forest Service Table from Baker and Votapka (1990) shows the diversity of spawning times of various fish likely to be encountered at culvert sites in North America.

These timings are not absolute and may vary according to local conditions such as temperature, etc. The time of migration may precede the spawning times in many cases, or it may be triggered by spatial needs, water temperature, or other factors. It

SPAWNING PERIOD
(Month)

is necessary for the biologist to specify the time of migration for all species required to be passed at a proposed or existing site as a first step in developing a solution to any potential problem. A second step is to classify, if possible, the species requiring passage into a category of swimming ability. Bell (1991) has used a four-part grouping or classification based on his own collection of swimming-speed data. The four groups are (I) small, weakly motivated fish, (II) poor swimmers, (III) medium-length fish, and (IV) large, highly motivated fish. This classification is devised to fit a table relating culvert velocities and lengths to swimming abilities, and one can readily see that it is very subjective in nature. At the other extreme, Alaskan biologists and engineers have developed a very detailed categorization for a single species, Arctic grayling, and have theorized from available data the swimming speeds of these fish under anaerobic and aerobic conditions (Behlke et al., 1991). They consider the Arctic grayling to be a weak swimmer and have developed a computer program to assist biologists and engineers in developing designs suitable for this classification.

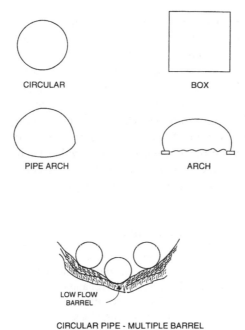

CIRCULAR   BOX

PIPE ARCH   ARCH

LOW FLOW
BARREL

CIRCULAR PIPE - MULTIPLE BARREL

Figure 7.1 Some types of steel and concrete culverts.

In addition to swimming ability the biologist must be prepared to provide the limits for timing of migration, so that it can be related to culvert design and timing of flood flows. It must be known or deduced what triggers the migration and whether a delay in migration can be imposed without danger to the fish, so that sudden extremes of flow will not adversely affect the fish. In other words, a sudden high flow lasting a few days will not necessitate a design to accommodate fish passage at high flows if the fish can be delayed for a few days without affecting their successful migration. Similarly, timing can be important in relation to temperature changes of the water, which Gould and Belford (1986) discuss in relation to trout passage in Montana. This factor of timing therefore is vital in the engineering design, which will be dealt with next.

7.3 ENGINEERING CONSIDERATIONS

When an engineer sets out to design a culvert, there are many factors to be taken into account. Among these are size and shape of the culvert, material from which it is made, the alignment, erosion that might take place, life of the structure, etc. Most important are the size, shape, and material of the culvert, some types of which are shown in Figure 7.1. The type selected depends on the maximum flow to be accommodated in the pipe, which is known as the design flow, and other factors such as the material of the fill, foundation conditions, etc. The engineer can arrive at the design flow in several ways. One is by analyzing the stream flow records and calculating the probability of a flood of a certain magnitude in the given life of the structure. This method is best, although it should not be used where stream flow records are sparse. Another method is by constructing a unit hydrograph and predicting the maximum. Once the design flow has been decided on, the engineer proceeds to make a decision on the type of pipe, the materials, and the other factors noted.

It is here that the data provided by the biologist must be taken into consideration. Reference to Appendix A, "Elementary Hydraulics," will show that the velocity of flow in an open channel and in a pipe is governed by the formula $V = 1.486/n\ R^{2/3}\ S^{1/2}$. This is known as Manning's formula, where R is the mean hydraulic radius in feet (the area of the cross section of water divided by the wetted perimeter), S is the gradient expressed as the drop in the channel divided by its length, and n is the roughness coefficient of the channel, some values of which are given in Appendix A. Additional values of use in culvert design are given in the following table from Kenney et al. (1992).

Surface	Description	Manning's n
Concrete pipe	Good joints, smooth walls	0.011–0.013
Concrete box culvert	Good joints, smooth finished walls	0.012–0.015
CMP,[a] annual corrugations	2⅔ × ½ in. corrugations	0.027–0.022
	6 × 1 in. corrugations	0.025–0.022
	5 × 1 in. corrugations	0.026–0.025
	3 × 1 in. corrugations	0.028–0.027
SPP,[b] annular corrugations	6 × 2 in. corrugations	0.035–0.033
	9 × 2½ in. corrugations	0.037–0.033
CMP, helical corrugations, full circular flow	2⅔ × ½ in. corrugations	0.012–0.024
Offset baffles (water depth 0.53 × culvert diameter)	Baffle height = 0.1 × diameter of culvert	0.047
Mountain stream channels	Gravel, cobble, few boulders	0.040–0.050
	Cobbles with large boulders	0.050–0.070

[a] CMP, Corrugated metal pipe.

[b] SPP, Structural plate pipe.

7.4 FINDING A SOLUTION FOR NEW CULVERTS

The biologist must now be in a position to add to the data on design flow the flow or flows at which fish passage should be assured (the "fish" flow). These extra data will be essential in deciding the pipe material (thus, setting n in the formula), the slope (thus, setting S in the formula), and the wetted perimeter R. These factors must be correlated so that the velocity V in the pipe will be within the swimming ability of the fish it is desired to pass.

There is a simple solution to this problem when the fish passage flow is comparatively low. In this case, the slope S is the same as the stream bed above and below the culvert. The materials that have been placed in the bottom of the culvert are the same as those in the stream bed and thus have the same n, and the hydraulic radius R remains the same as that of the stream bed. Thus the velocity V is the same as the stream, and fish that are able to ascend the stream can readily pass through the culvert.

The primary rule in the design of new culverts is this: the stream bed through the culvert must be as close as possible to the original stream bed in slope, bed material, and wetted perimeter.

It is not always possible to meet these idealized conditions, and the most common departure from it is that the "fish" flow is too high to maintain the channel bottom roughness n and a constant hydraulic radius R through the pipe. Thus the velocities must increase, and if they increase too much the fish cannot pass through. One way of compensating for this is by increasing the roughness n of the pipe. It is best to use pipe with annular corrugations for this, and the deeper the corrugations the better for roughness and thus for fish passage. The design engineer, however, may wish to avoid such pipes because increasing n limits the flood flow that the pipe can accommodate and thus makes it necessary to use a larger and more expensive pipe. This is one of the compromises that will have to be resolved to ensure that the culvert is passable to fish.

Another way to make the culvert passable to fish at higher flows is to attach a series of baffles within the pipe. These have an effect that is similar to but more intensive than that of corrugations, but they also reduce the level of flood flows that the pipe can accommodate. Baffles are not recommended in designs of new culverts if another solution can be found. They can be useful for improving existing culverts, however, and will be discussed in more detail later.

Another weakness in design of culverts for fish passage is that a single large pipe often permits the flow to spread out at low flows to the point where there is insufficient depth for the fish to swim. This can be overcome to a certain extent, especially in cases where the site lends itself to this method, by installing multiple barrel culverts. In this case one barrel of the culvert should be placed lower than the others (see Figure 7.1) so that the flow is concentrated in one pipe and there is sufficient depth for fish to swim at low flows. It is also a problem at box culverts, which have a smooth, flat bottom that is often impassable for fish. In this case multiple culverts can also be used, or a section of the floor of a single culvert can be lowered to concentrate the flow at low flows.

A third problem can arise after the culvert is built, unless great care is taken in design. Erosion can take place at or below the culvert outlet. Since the culvert is designed to take the maximum flow in the stream, this means that at flood flows the pipe will be running full and there may even be considerable head on the intake end. The result is extremely high velocities in the pipe with inevitable scouring resulting below the outlet. This must be foreseen in the design, and sufficient riprap placed below the outlet, or a formal concrete stilling basin designed to prevent serious scour and leave the outlet passable to fish. Many of the species encountered at culvert sites are not jumpers and cannot ascend a drop of more than a few inches. Thus it is necessary to either pave the outlet with riprap and river gravel thus allowing the natural stream bed to resist erosion, or construct a stilling basin of concrete or riprap or both to a design that will permit the fish to ascend and will remain in place after several years of flood flows. Figure 7.2 shows this problem, which is known as "perching," and suggests a type of solution.

It will be seen that culvert design to include fish passage frequently results in a conflict of interest between the highway engineer and the biologist.

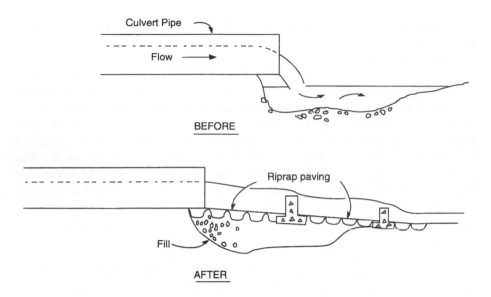

Figure 7.2 Two elevations showing an extreme case of perching corrected by concrete weirs and riprap.

The highway engineer is interested in getting by with the smallest and smoothest possible pipe to carry the design flow at the lowest cost, while the biologist wants to increase the roughness n of the pipe, which will in turn decrease its capacity. In addition the biologist requires a formal addition to the structure to prevent scour which adds to the cost and may not be required for safety of the structure. Nevertheless, the biologist must insist on measures that will provide passage for the fish, and try to work out a solution that involves the least additional cost and yet provides satisfactory fish-passage conditions.

7.5 MAKING EXISTING CULVERTS PASSABLE

This is treated as a separate problem because so many culverts have been built without attention to a potential problem of fish passage, and thus a slightly different approach is called for.

Many existing culverts have the condition known as perching, that is, excessive scour below the downstream end of the pipe, which leaves a drop of some magnitude that fish are unable to ascend. These are usually associated with pipe culverts, as shown in Figure 7.2. So this becomes one of the main problems with existing culverts. There have been many solutions used for such culverts, all of which are of the same principle. It is necessary to bring the tailwater to a level that makes it possible for fish to ascend. This has been done by installing permanent walls or baffles across the scour pool composed of riprap or concrete as described earlier, creating a series of steps the fish can ascend. These walls or baffles can be notched, depending on the needs of the fish or the structure in order to ensure permanency.

The other problem is one of high-velocity flow within the culvert. There are many ways of reducing the velocity at fish migration flows, all of which involve placing baffles within the culvert. These baffles can consist of concrete, steel, wood, or other material, or they may simply be large rocks anchored to the bottom at intervals. Each culvert will have unique properties that make one of these systems more appropriate than the others. A few representative examples are shown in Figures 7.3 and 7.4.

The offset baffle shown in Figure 7.3, originally described in McKinley and Webb (1956), can be constructed of concrete, steel, or wood, and fastened to the culvert walls by bolts, or grouted. This type of baffle is unfortunately subject to sedimentation, and therefore allowance must be made for annual maintenance. It was found during hydraulic tests that doubling the standard height shown or spacing the baffles closer than shown increased the resistance and therefore improved the n in Manning's formula. Also it was found that they worked best when the fish-passage flow just overtops them, so care must be taken when considering them that the fish-passage flow is in this range. If not, alteration of the height and spacing of baffles can be considered, as can the use of other baffles such as those shown in the diagrams.

The slotted weir baffle shown was found to be as effective as the offset baffle for fish passage; it also is less complicated in design and construction. However, weirs without the slot were impassable to nonleaping fish, particularly at low flows. Weir baffles tend to retain more bedload, but are less subject to jamming by brush and floating debris. Weirs of either the slotted or nonslotted type also need frequent maintenance.

Spoiler baffles are the most expensive to design and construct, but are effective in velocity reduction, and probably have less tendency to gather bedload and floating debris. It was found that metal plate spoiler baffles were effective in reducing velocities and were simple to construct.

The reinforcing bar and boulder structure shown in Figure 7.4(A), as installed in Alaska, is effective for fish passage, particularly at low flows, provided that it does not get clogged with debris to the point where water flows are impeded at one or more points, creating concentrated drops in the water-surface profile. For this reason frequent maintenance is a necessity.

Finally, the angle iron and rebar type of baffle shown in Figure 7.4(B), used in Montana for trout passage, has been quite successful for steep gradients, and is easily maintained by removing the entire frame from the culvert.

All these baffles and more are described and summarized in Kenney et al. (1992), and references are given there as to the origin of each type. Readers requiring more detail would be well advised to consult this reference first, and select references for further study from it.

All these solutions involve problems to the culvert owner. They lower the capacity of the pipe, thus arbitrarily reducing the design flow. The pipe hydraulics will have to be reviewed under the new baffled conditions, and if necessary a hydraulic model study of the proposed alterations should be made to determine the new value of n and the resulting effect on the design flow.

ELEVATION

PLAN

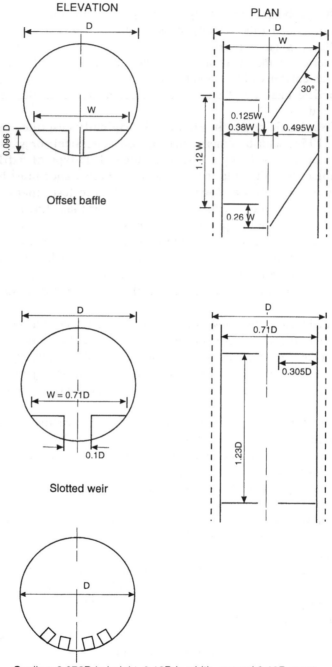

Offset baffle

Slotted weir

Spoilers 0.076D in height, 0.10D in width, spaced 0.10D apart.
Alternate rows of 4 and 3 placed at 0.46D intervals

Spoiler baffle design

Figure 7.3   Three types of culvert baffles from various sources as summarized by Kenny et al. (1992).

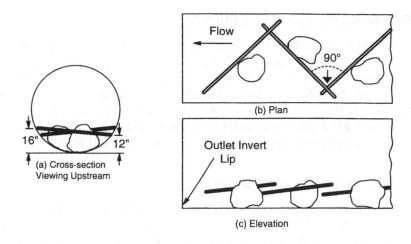

A. A type of channel roughening retrofitted in a culvert in Alaska

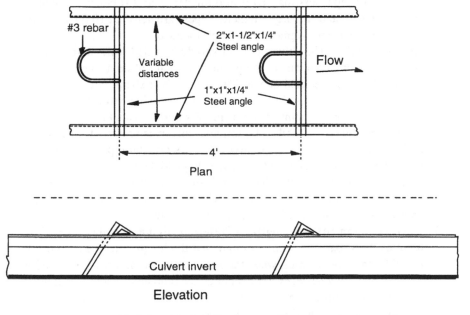

B. A bedload collector used to increase channel roughness in culverts in Montana

Figure 7.4 **Two methods of retrofitting used in culverts to increase roughness: (A) A type of channel roughening retrofitted in a culvert in Alaska; (B) a bedload collector used to increase channel roughness in culverts in Montana.**

7.6 RECOMMENDATIONS — GENERAL

Following is a summary of recommendations made by the many sources noted in the references. The summary is not complete but includes those which the writer considers most important. There are many other publications quoted in the references cited, and interested readers are well advised to consult those that seem appropriate to their problem. However, these recommendations are considered to be a fair cross section of state and federal government views in the U.S. and Canada. Where they refer mainly to a particular species or group of fish, this fact is noted.

7.6.1 For New Culvert Installations

1. **Culvert gradient** — It is generally agreed that this should be no steeper than the stream bed gradient above and below the culvert. Some reports advocate less than this, even a flat gradient, but my objection is that a departure from the original stream gradient upsets the natural regime of the stream and should be avoided.
2. **Culvert elevation** — It is generally agreed that the bottom of the culvert should be depressed below the normal stream bed gradient by a certain amount. For pipes less than 10 ft in diameter, 1 to 2 ft below gradient is recommended, and for large-diameter pipes (greater than 10 ft), one fifth of the diameter below gradient is recommended.
3. **Bed material** — The culvert bottom should be filled with natural stream bed material or equivalent, to bring it up to the natural stream bed elevation and gradient as specified above. The stream bed shape in cross section should be duplicated as closely as possible to provide sufficient depth for fish to swim and to maintain the original wetted perimeter.
4. **Culvert material** — While a bottomless culvert is preferred, such as the arch culvert in Figure 7.1, the next choice is a metal pipe with corrugations. The corrugations would be such as to yield the maximum coefficient *n* in the Manning formula. The least desirable choice is a box culvert of concrete, which provides a wide, level (in cross section) channel that is extremely difficult for fish to ascend at low flows.
5. **Culvert alignment and length** — Preferred alignment is the same alignment as the stream, but keep in mind that the length must be as short as possible. In extreme cases, when the angle between the stream and road is very small, it may be necessary to change the road alignment to meet these criteria. Such drastic measures will depend on the value of the fish and their demonstrated need to ascend the stream.
6. **Culvert erosion** — Potential erosion at both the culvert inlet and outlet must be avoided. This can be done by armoring the stream bed at inlet and outlet with riprap, concrete, or other materials, in order to avoid potential drops in water surface at each end of the pipe.
7. **Culvert maintenance** — Arrangements must be made for regular inspection and maintenance of culverts, as with any artificial structure placed in a stream in the migratory path of fish.

7.6.2 For Retrofitting of Old Culverts

The method of making an existing culvert passable to fish will depend on the conditions that render it impassable. The two most common conditions are perching, which prevents the fish from entering the culvert, and a velocity of flow in the pipe that is too high for the fish that enter the pipe to complete their ascent through it.

1. **Perching** — This was dealt with in some detail in Section 7.5. Placing riprap or concrete sills or baffles usually provides the answer. The extent of erosion is known with existing culverts, so a better solution can often be provided than with new culverts. Paving of the eroded basin, and if necessary the placing of concrete sills or baffles, notched as necessary, can provide an answer. This will definitely be needed where the invert of the culvert pipe has been placed at or above the natural stream bed, and a small fishway is necessary to provide access for the fish.
2. **Too high a velocity in culvert** — If this condition occurs, and filling the bottom with bed material does not solve the problem, it may be necessary to provide baffles in the culvert. Many different types of baffles, ranging from fastening boulders at intervals along the bottom of the culvert to placing formal concrete baffles, have been used. In the case of a minor velocity obstruction to ascending fish the boulders are usually sufficient. It is best to have some means of holding them in place, such as by cementing them to the culvert walls or bottom, or by holding them with steel reinforcing rods or bolts drilled into the pipe, or with angle irons fastened to the pipe.

7.7 LITERATURE CITED

Baker, C.O. and F.E. Votapka, 1990. Fish Passage Through Culverts, Prepared by U.S. Dept. Agric., Forest Serv. U.S. Dept. Transp., Fed. Highway Adm. Rep. No. FHWA-FL-90–006.

Behlke, C.E., D.L. Kane, R.F. McLean, and M.D. Travis, 1991. Fundamentals of Culvert Design for Passage of Weak-Swimming Fish, Prepared for Alaska Dept. Transp. in cooperation with U.S. Dept. Transp., Fed. Highway Adm. Rep. No. FHWA-AK-RD-90-10.

Bell, M.C., 1991. Fisheries Handbook of Engineering Requirements and Biological Criteria, Prepared for Fish Passage Dev. Eval. Prog., U.S. Army Corps of Engineers, North Pac. Div., Portland, OR.

Gould, W.R. and D.A. Belford, 1986. Prepared by Montana Coop. Fish. Res. Unit, Montana State Univ., for Montana Dept. Highways, Proj. 8093.

Kenney, D.R., M.C. Odom, and R.P. Morgan, 1992. Blockage to Fish Passage Caused by the Installation/Maintenance of Highway Culverts, prepared by the Appalachian Environmental Lab., Univ. Maryland for State Highway Admin., Maryland Dept. of Transp.

McKinley, W.R. and R.D. Webb, 1956. A Proposed Correction of Migratory Fish Problems at Box Culverts, Washington Dept. Fish., Fish. Res. Pap. No. 1(4).

7.8 OTHER REFERENCES

Browning, M.C., 1990. Oregon culvert fish passage survey. Prepared for Western Federal Lands, Highway Division, Federal Highway Administration.

Katopodis, C., P.R. Robinson, and B.G. Sutherland, 1978. A Study of Model and Prototype Culvert Baffling for Fish Passage, Canadian Fish. Mar. Serv. Tech. Rep. No. 828.

Normann, J.M., R.J. Houghtalen, and W.J. Johnston, 1985. Hydraulic design of highway culverts. Report No. FHWA-IP-85–15. U.S. Dept. of Transportation, Federal Highway Administration.

Rajaratnam, N.C., C. Katopodis, and N. McQuitty, 1989. Hydraulics of culvert fishways II: slotted-weir culvert fishways. Can. J. Civ. Eng., 16, pp. 375–383.

Rajaratnam, N.C., C. Katopodis, and S. Lodewyk, 1989. An Experimental Study of Fishways with Spoiler Baffles, Tech. Rep. No. WRE-89–4. Dept. Civ. Eng., Univ. Alberta.

ADDENDUM

The manuscript for this second edition was submitted for publication in February 1990, but its writing was completed during the previous year. It will be noted that most of the cost data is updated to the late 1980s. During the ensuing long delay in publication progress has been made on many fronts in research and application in the design of the type of facilities the volume covers. Rather than painstakingly amending the text to incorporate the details of this progress, I have chosen to bring the important advances to the reader's attention in this addendum, with the hope that they might be found useful to biologists and engineers who are interested in the specialized fields noted.

The three fields in which there has been notable progress are as follows:

1. Improvements to facilities on Columbia River dams
2. Improvements to revolving screens
3. Fishways for catadromous fish in Australia

Details of these follow.

AD.1 IMPROVEMENTS TO FACILITIES ON COLUMBIA RIVER DAMS

First, the change from orifice weirs to vertical slot baffles to control the flow to many Columbia River fishways was drawn to my attention by Corps of Engineers biologists. This change was made in stages over recent years mainly to facilitate passage of American shad. As is widely known, mature, migrating American shad exhibit behavior characteristics that are unique, and this change allows them to migrate near the surface, which they prefer, while still fulfilling the original purpose of applying a measure of control of flow to downstream baffles of the fishway.

Second, the use of gravity flow for attraction water at some of the lower Columbia dams has been changed by introducing auxiliary generating facilities to dissipate the energy in the auxiliary water prior to its passing into the fishway upstream of the entrances. This is a normal progression one would expect now that hydroelectric power is becoming higher priced and in shorter supply.

Finally, the attempts to provide safer passage to downstream migrant fish on the Columbia River continue to receive much attention by biologists. By increasing the size of the revolving screens (doubling their depth) in the powerhouse intakes, it is hoped to increase their efficiency of diversion to 80 to 90%. Further refinements to the bypass facilities, as well as capturing and transporting of fish from upstream dams past several downstream dams, have also shown some success in solving the downstream migrant problem.

The net result of all these ongoing activities to improve fish passage has proven to be less than one could wish for in conserving the Columbia River salmon runs. Nevertheless the upstream facilities, as described in the text, remain a model of the best in fish-passage design, and are well worth describing in detail in this revised edition.

AD.2 IMPROVEMENTS TO REVOLVING SCREENS

Improvements to the link belt type (also known as the Ristroph type) of revolving screens have recently been made as a result of work supported by the Hudson River Fishermens Association and several power companies in the New York region. Fletcher (1990) describes experiments on the screens utilizing the improvements noted in Figure 6.12 and described in the accompanying text, which demonstrated that even with these improvements, in the situation described, the screens were showing unacceptably high losses of striped bass, white perch, Atlantic tomcod and pumpkinseed, of about 5 to 7 cm in length.

The screens were studied in a hydraulics laboratory, and it was found that by adding an auxiliary screen to the lip of the trough, and making further minor adjustments in the trough, as shown in Figure AD.1, hydraulic conditions were altered sufficiently favorably to reduce losses from 53 to 9% for striped bass at approach velocities of 30 and 45 cm/sec. Other juvenile fish had a comparable reduction in percentage loss.

The work was done in a classical manner with a biologist working with hydraulic engineers to improve the screens. The resulting improvements in losses for the fish noted under the conditions described is most commendable.

AD.3 FISHWAYS FOR CATADROMOUS FISH IN AUSTRALIA

As mentioned in the main text, both South Africa and Australia have young fish that migrate upstream after being spawned in the estuary of the river. In New South Wales, Australia, this causes an acute migration problem, since there are many low dams in the rivers for irrigation and other purposes in the lowland areas along the coast. Some of these dams included fishways, which were based on early European or North American designs for salmonids, and only a few of the catadromous fish have been able to surmount them. In searching for a solution to this problem, Mallen-Cooper (1992) describes the testing of a low-head vertical slot fishway that successfully passed juvenile Australian bass and barramundi. The fish ranged in length from 40 to 93 mm for the bass, and ~43 mm for the barramundi.

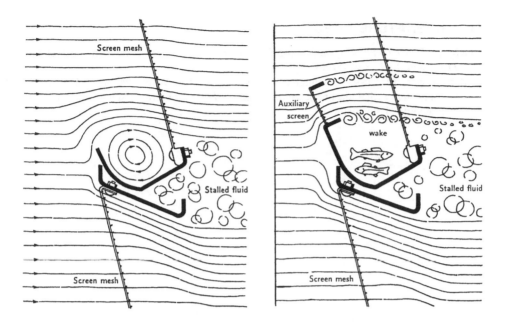

Figure AD.1 **Two sectional views of adjoining panels of traveling screen showing (left) the extended lip as recommended in Figure 6.12, and (right) further extension to the lip in the form of an auxiliary screen, the addition of which greatly improved the efficiency of diversion. (From Fletcher, R.I., 1990.** *Trans. Am. Fish. Soc.,* **119(3), pp. 319–415. With permission.)**

The fishway was based on the Seton Creek model shown in Figure 3.13 of the main text, scaled down to 1 m in width, with the other dimensions reduced proportionately. It was constructed in a hydraulics laboratory in Manly near Sydney, and actively migrating fish were taken from rivers in the vicinity for the tests. One baffle of the three in the fishway was carefully controlled, and maximum velocities were recorded. It was assumed that fish that could pass through could ascend a series of baffles (how many is unknown at present) operating at this maximum velocity. The fish ascended successfully at velocities of 1.0 to 1.4 m/s, which corresponds to a head per baffle of 50 to 100 mm, or about 2 to 4 in. Generally 90 to 100% of the fish successfully passed upstream through this short vertical slot fishway at these maximum velocities.

These tests were admittedly lacking in many respects, mainly because they were not conducted in the field under natural conditions. However they led to the installation of several new fishways of identical dimensions, and counts are being made on these to establish the degree of success under field conditions. Much work remains, for example, to establish the maximum velocities allowable for fishways for a number of species of various sizes, but this is considered to be a good start to solution of a previously unsolved problem.

It is hoped that with these notes I have brought the subject matter of design of fishways to a reasonably up-to-date state.

AD.4 LITERATURE CITED

Fletcher, R.I., 1990. Flow dynamics and fish recovery experiments: water intake systems, *Trans. Am. Fish. Soc.,* 119(3), pp. 393–415.

Mallen-Cooper, M., 1992. Swimming ability of juvenile Australian bass, *Macquaria novemaculeata* (Steindachner), and barramundi, *Lates calcarifer* (Bloch), in an experimental vertical-slot fishway.

APPENDIX A
ELEMENTARY HYDRAULICS

A.1 INTRODUCTION

In G.E. Russell's (1940) text entitled *Hydraulics,* the author defines hydraulics as "that branch of mechanics which deals with the laws governing the behaviour of water and other liquids in the states of rest and motion."

Many species of fish spend all or part of their lives in freshwater, so that the laws governing the behavior of water can be quite important to an understanding of the environment of these fish. Such an understanding is essential in managing a fishery depending on stocks of anadromous fish or fish that spend all their lives in freshwater.

One of the fundamental phases of the study of hydraulics is the development of methods for measuring water flow in volume per unit of time. In fisheries management such measurements are often the basis for determining what populations of fish a given stream will support, and what changes will take place in the fish populations with predictable changes in volume of flow per unit of time.

As an example let us consider two rivers that have the same width and depth, but an entirely different volume of flow per unit of time. This is possible with rivers of different gradient and bottom characteristics. It is likely that the difference in quantity rate of flow will mean that the two rivers support quite different fish populations. It could also mean that the diversion of a certain quantity of water from one will have quite a different effect on its fish population than would the diversion of an equal quantity from the other.

Fortunately, stream flow measurement is not complicated or difficult to understand, and it is hoped that the readers, regardless of their academic training, will be able to understand readily the following explanations of this and other phases of hydraulics.

While the ability to measure fluids in motion is very useful, other phases of hydraulics can also be useful in fisheries management. A knowledge of the mechanics of flow through orifices is important in designing fishways, particularly those with submerged orifice baffles. The mechanics of flow over weirs is of equal importance in designing fishways with weir type baffles. Flow in open channels has also been studied for many years and the knowledge gained from these studies can be useful in designing open channels for various fish cultural activities, and in studying the natural spawning grounds of salmon and trout. Care must be exercised, however, in applying the laws of open channel flow to natural streams, as will be seen in the section of the text dealing with this subject.

A.2 UNITS OF MEASUREMENT

Fortunately the common units of measurement used in hydraulics are few in number. In the English system volumes are expressed in gallons (gal), cubic feet (ft³) or acre-feet (acre-ft); velocities in feet per second (ft/sec); and volumes per unit of time in cubic feet per second (cfs) or gallons per minute (gpm).

In the metric system volumes are usually expressed in cubic meters (m³), velocities in meters per second (m/s), and volumes per unit of time in cubic meters per second (m³/sec).

Gallons may be either Imperial or U.S. measure, and since these differ, care must be taken to ensure that the correct measure is being used. An Imperial gallon of water contains 277.27 in.³ and weighs 10 lb. A U.S. gallon of water contains 231 in.³ and weighs 8.34 lb. Because the gallon is a comparatively small unit of measurement, it is never used to describe stream flow, but is used extensively to describe pump capacity and domestic water use. If it is necessary to convert frequently from Imperial measure to U.S. measure, a convenient equivalent is 1 Imperial gallon equals 1.2 U.S. gallons.

The cubic foot is the most convenient unit for measuring small volumes of water because it readily converts to the normal units for expressing flowing water (in volume per unit of time) in pipes, open channels, and natural streams.

The acre foot is the most convenient unit for measuring large volumes of water such as the capacity of large reservoirs and impoundments.

While volumes per unit of time can be expressed in any of the foregoing units, the most useful expression is that combining cubic feet and the time interval of one second, or cubic feet per second. This is abbreviated cfs, but is sometimes expressed for convenience as second-feet (American usage), or as cusecs (British usage). The abbreviation cfs has been used throughout this text.

One cfs is a comparatively large unit, so that a smaller unit is more convenient in describing pump capacity, or flows for domestic use or hatchery ponds. The common units of measurement used for these purposes are gallons per minute (gpm) or millions of gallons per day (mgpd).

A.3 DEFINITIONS

Many formulas have been developed both theoretically and empirically for solving problems in hydraulics. Only the simpler, more commonly used formulas will be explained in this text. Readers are referred to the many excellent texts on hydraulics, some of which are listed in the references at the end of this appendix, if they wish to go into the subject in more detail.

All the terms used in the formulas listed in this text are defined below. Care should be taken in attempting to apply these definitions to formulas taken from other texts, because different texts often use different nomenclature.

Q = discharge in cubic feet per second or any of the other units expressing volumes per unit of time defined in previous sections

A = cross-sectional area in square feet or other convenient unit

V = average or mean velocity in feet per second or other convenient unit

g = acceleration of gravity (usually considered to be 32.2 ft/sec/s)

H = head in feet acting on a weir

h = head in feet acting on an orifice, and also velocity head ($V^2/2g$)

C = coefficient of discharge (dimensionless) for an orifice or weir, or coefficient of roughness for an open channel or pipe

R = hydraulic radius of a stream in feet, which is equal to a cross-sectional area *(A)* divided by the wetted perimeter* of the cross section

S = gradient or slope of open channel expressed as drop in feet divided by the length of the channel in feet over which the drop takes place (assuming total energy gradient, slope of water surface, and grade of channel are the same)

n = coefficient of roughness used in the Manning formula for open channels or pipes

L = length of weir crest in feet

* Wetted perimeter is defined in Section A.8.

Care should be taken in applying formulas to the solution of hydraulic problems to ensure that the units of measurement used in any one formula are consistent. In addition, because some of the formulas are empirical, they will yield an inaccurate or at best only an approximate answer if the conditions under which the formula was derived have not been duplicated. If caution is used in their application, however, they should be sufficiently reliable to be useful for a wide range of fisheries problems.

A.4 ASSUMPTIONS

It is believed desirable to state here at the outset what academic training or knowledge is necessary in order to understand and be able to apply the material included in this appendix. It is assumed that the reader has a knowledge of basic mathematics and physics up to a senior matriculation of first-year university level. This should include an understanding of logarithms, which will be required to solve some of the formulas.

A knowledge of calculus would be required to derive some of the formulas used and to follow the general subject in more detail in other texts, but is not required to understand or make use of the material in this appendix. In addition it is not necessary to have prior engineering training. This section of the text is intended primarily for the use of biologists and fisheries management technicians.

A.5 FUNDAMENTAL EQUATION OF FLOW, $Q = AV$

The most frequently used formula in applications of hydraulics is

$$Q = AV \qquad (1)$$

Assuming that A is the cross-sectional area of a pipe in square feet, with the pipe running full of water, and V is the average velocity of the water over the cross section, then it can be seen that the quantity Q flowing through the pipe per unit of time is the product of A and V. In accordance with the explanation in the section on units of measure, Q would be expressed in cubic feet per second (cfs).

There are many examples of the use of this equation in hydraulics: to measure quantities of water used for irrigation, domestic consumption, and industry; and to measure stream flow available for these uses. These will be detailed in a later section on practical applications of measurement of flowing water.

It should be noted at this point that while A is a comparatively straightforward quantity to measure or calculate, V is not. Many methods have been developed for measuring V, some of which will be explained in the later section. Many empirical methods and formulas have also been developed for calculating V, all of which have the weaknesses pointed out previously in the section on definitions. Anyone who has thrown pieces of wood into a stream and noticed how those pieces in the center of the stream travel much faster than the ones dropped near the margin, will realize what a difficult task it is to determine the average velocity of the stream at any one cross section. The same difficulties are apparent in the determination of velocity in other conveyors of water such as pipes and flumes.

A.6 FLOW THROUGH ORIFICES

An orifice can be defined as any opening with closed perimeter through which a fluid flows. Theoretically, water discharging through an orifice under a head h has a mean velocity equal to that acquired by a body falling freely in a vacuum through a distance equal to h. This velocity is equal to the square root of the product of the head h and twice the acceleration of gravity g. Theoretically therefore, the formula would be

$$V = \sqrt{2gh} \qquad (2)$$

However, the actual mean velocity is never equal to the theoretical, and must therefore be reduced by a multiplying factor or coefficient which is less than unity.

Because $Q = AV$, and we have an equation for expressing V through an orifice, we could determine Q for the orifice by substitution in our original equation. We find, however, that the cross-sectional area of the water passing through the orifice is never equal to the area of the orifice, and we must use another coefficient to correct for this. It is possible to combine the two coefficients (one for area and one for velocity) into one, and write the formula as follows:

$$Q = CA\sqrt{2gh} \qquad (3)$$

The coefficient C is then known as the *coefficient of discharge*, which varies considerably for different edge shapes of orifices. If the orifice does not discharge freely into the air but is submerged, it is known as a submerged orifice and the head h used in the formula is the difference between the water levels on both sides of the

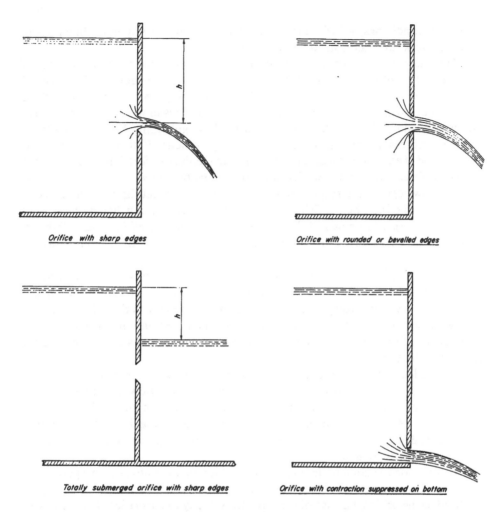

Figure A.1 **Four orifices with the same diameter, showing the possible differences in three of them in contraction of the jet, which affect the coefficients of discharge *C*. The fourth (lower left) shows a case where the orifice is submerged, and the contraction is similar to the example immediately above.**

orifice. Submergence does not appear to change the coefficient of discharge *C* significantly, however, the main factor affecting *C* is the shape of the edge of the orifice itself. Figure A.1 illustrates several different edge conditions, and shows how the jet is contracted to different degrees depending on the shape of the edges. Contraction may also be partly suppressed, as shown for the orifice at the floor of the container (lower right, in Figure A.1). In general, the sharp edge produces the lowest value of the coefficient *C* while the rounded edge produces the highest, which is close to unity.

Because it is easily constructed and calibrated, the orifice provides a convenient method for measuring flows. As a result, much experimental work has been done on determining values of *C* for certain specific types of orifices that are used for measuring purposes in agriculture and industry. Many of these can be found listed in various handbooks of hydraulics. While these are not very often of direct use in fisheries work, they are sometimes applicable.

Unfortunately, not very much experimental work has been done on the types of orifices that are of direct practical interest to fisheries conservationists. Examples of these are gates at dams and water-control works of all types placed in the path of upstream or downstream migration of fish, and the orifices in weir type, orifice type, and vertical slot fishways. A knowledge of the laws governing flow through these types of orifices is important for the following reasons. If there is a sufficient background of experimental data to enable one to predict the coefficient of discharge with confidence, the mean velocity in the jet can be predicted.

A knowledge of the magnitude of the velocity to expect through an orifice is helpful in predicting whether fish can ascend through it, assuming that the swimming ability of the fish is known. This applies to submerged gates in a dam, fishway baffles, or any structure that fish are intended to pass through or might attempt to pass through.

Conversely, it is often the practice to lead young fish migrating downstream away from the face of a screen or other device such as a louver by means of a bypass, the entrance to which frequently takes the form of an orifice. The flow characteristics of the orifice in this case are extremely important, because the velocities and acceleration of the flow through it are the most important factors in determining the success of the bypass.

Before leaving the subject of orifices, some of the factors other than the shape of the edge of the orifice that might affect the calculation of flow through the orifice should be noted. Recognition of these factors is extremely important because they can lead to a completely erroneous result if they are not recognized and due allowance made. Briefly, these factors are divided into two groups, those affecting velocity, and those affecting head.

If the orifice is situated in a channel in which there is already an appreciable water velocity, the actual velocity through the orifice will be a function not only of the head on the orifice, but also of the approach velocity to the orifice. The result will be a more complex relationship to include this new variable in addition to the head, and more than likely it can be solved only by duplicating the actual conditions on an experimental basis. However, if a significant so-called *velocity of approach* exists and is recognized, the reader can be forewarned that the simple methods of calculation outlined here might lead to an erroneous answer, and if importance is attached to the result, the advice of a competent hydraulic engineer should be sought.

Similarly, erroneous calculations can result from improper measurement or calculation of the head acting on the orifice. It may be that the orifice drains a reservoir or some other static or partly static body of water. In this case it is possible for the head to decrease significantly as the reservoir is emptied, leading to a progressive decrease in the discharge and the velocity through the orifice as the water surface lowers. Another instance of a situation where an erroneous calculation might result is where flow is through a submerged orifice such as a gate, particularly if the water is passing through at a high velocity. In such a situation the reader may note a standing wave or *hydraulic jump* some distance below the orifice. In such cases the simple formula previously outlined will not yield accurate results.

A.7 FLOW OVER WEIRS

In hydraulics a *weir* is generally considered to be that part of the crest of a barrier placed across an open channel over which the water discharges. This overflow part of the crest can extend the full length of the barrier, that is, from bank to bank of an open channel, or it can be shaped in the form of a notch in any one of a number of geometrical shapes. The term *weir* is often applied to the whole structure of the barrier, a practice that probably originated with weirs that stretched completely across open channels. The term is also applied to fences or racks made of wire screen, netting, or pickets, for trapping and counting migratory fish. However, in this appendix the term is used only in the hydraulic sense.

Like orifices, weirs offer a good, practical method of measuring rate of flow of water, and as a consequence much experimental work has been done to perfect formulas applicable to a number of different shapes of weirs. These usually have been classified according to the shape of notch used, such as *rectangular, triangular* (V-notch), and *trapezoidal,* and also according to the shape of crest such as *sharp-crested* and *broad-crested.*

It is not the intention here to discuss the various formulas that have been derived because they represent refinements beyond the scope of this text. We will present here one simple formula that can be used to approximate discharges under most conditions and indicate the possible sources of error in its use.

The discharge over a weir can be calculated from the formula

$$Q = 3.33LH^{3/2} \qquad (4)$$

This is the formula for a sharp-crested rectangular weir without velocity of approach. The Figure 3.33 is actually a coefficient that applies to this type of weir. The coefficient can vary from 2.6 for weirs of very broad crest (as opposed to a sharp crest) to more than 3.33 for sharp-crested weirs under special conditions. However, the value 3.33 will give good approximate results for most weirs encountered in fisheries work. The exceptional cases where it does not give good approximations will be for broad-crested weirs (1) with a breadth of crest greater than 4 or 5 ft, or (2) with less than about 1 ft of head on the weir. Formula (4) cannot be applied to V-notch or trapezoidal weirs, either.

The most likely source of error in calculating weir flows is in neglecting to take the velocity of approach into account. Because many weirs that fisheries workers might have reason to study will have large velocities of approach, it might be as well to outline a simple process by which an allowance can be made for this factor. Formula (2), the theoretical formula for velocity of falling water, stated that $V = \sqrt{2gh}$. This can be squared and transposed to read

$$h = \frac{V^2}{2g} \qquad (5)$$

If the mean velocity of approach to the weir is known, this equation can be solved for h quite readily. In so doing, one is obtaining a theoretical value of the potential energy or head equivalent to the velocity energy or head that the water possesses in approaching the weir. By combining formulas (4) and (5), due allowance can be made for the extra head resulting from the velocity of approach, and the formula now becomes

$$Q = 3.33L\left[(H+h)^{3/2} - h^{3/2}\right] \qquad (6)$$

This formula should be used where the velocity of approach is high enough to make a difference in flow Q of sufficient importance to the user. It will be recognized that approach velocities as low as 1 ft/sec will make little difference in the answer, whereas velocities as high as 8 ft/sec will make a significant difference, as illustrated in the following example.

A low weir with a sharp crest is constructed across a fast-moving river. If the head H acting on the weir at a certain river stage is only 2 ft, the volume of flow

$$Q = 3.33 \times 1 \times 2^{3/2}$$

$$= 9.3 \text{ cfs/ft of length } L$$

If, however, there is a velocity in the river approaching the weir of 8 ft/sec, the volume of flow becomes

$$Q = 3.33 \times 1\left[(2+1)^{3/2} - 1^{3/2}\right]$$

$$= 14.0 \text{ cfs/ft of length } L$$

The depth of flow at the weir crest itself is likely to be somewhat less than H. (This is illustrated in Figure A.2.) It is therefore important that H be measured at a distance of at least 2.5 H upstream of the weir if accuracy is desired.

Other possible sources of error in computing the flow over weirs are end contractions and ventilation of the nappe. If the length of the weir crest is not large, it can readily be seen that the sides of the notch will have a great influence on the total flow over the weir. The extreme case is the V-notch weir, in which the crest itself is eliminated and only the sides control the flow. In these cases special formulas have been devised and the reader is referred to a hydraulics text or handbook for further details.

The condition of the nappe at any weir can have a considerable influence on the relationship between H and Q. The nappe is the sheet or jet of water that forms after the water has passed over the weir crest. It can be contracted after leaving the crest of the weir as in Figure A.2 or it might cling to the downstream face of the weir, with no air underneath. While these two conditions can produce significant differences in calculating flows, generally speaking they need not be considered by the fisheries biologist or technician. The condition of the nappe can be of importance to fisheries workers, however, in quite another way. As has been described earlier in this text, it has been found in some of the longer weir type fishways on the Columbia River that with certain dimensions of weir and pool, and under certain flow conditions, a wave action that oscillates from side to side has been set up in the fishway as a result of the nappe's alternately clinging to the downstream face of the weir and then springing free from the face. The resulting waves have reached several feet in height,

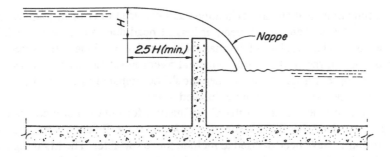

Figure A.2 Typical flow pattern over a weir with ventilated nappe.

overtopping the fishway walls. The methods of alleviating or avoiding this condition were discussed in an earlier section of this text.

One other condition known as submergence should be mentioned before leaving the subject of weirs. This is the case when the water surface downstream from the weir is above the level of the weir crest. This is frequently the case at low dams, particularly when flows are large enough to raise river levels downstream to the level of the weir crest. Once again this is a special case requiring detailed study beyond the scope of this text. However, the condition is noted so that it will be recognized when encountered and erroneous conclusions avoided.

A.8 FLOW IN OPEN CHANNELS

The study of flow in open channels is of extreme importance to the fisheries worker because it can be applied directly to the streams that are the natural environment of the fish. Unfortunately, formulas developed for artificial channels of uniform slope and regular cross section *must be applied with caution* to the natural channels of rivers, which are irregular in cross section and slope. A knowledge of these formulas, however, will help the reader to realize the complexity of the factors affecting flow in natural streams, and can be of direct use in the design of such things as artificial spawning channels.

The most useful formula for determining the flow to be expected in a given channel is one devised by Chezy nearly 200 years ago. It involves an expression for the velocity when the channel cross section, slope, roughness, and depth of flow are known, which can be stated as follows:

$$V = C\sqrt{RS} \qquad (7)$$

The value of V can be substituted in formula (1) to give the volume rate of flow Q. Although the terms of this formula have been previously defined, they are elaborated on here again for purposes of clarity. Here, R is the mean hydraulic radius in feet, which is defined as the area A of the cross section of the water in the reach under study, divided by the wetted perimeter. The wetted perimeter is the length of the line of contact of the water with the solid boundary of the channel cross section. The term S has been defined as the gradient or slope of the channel expressed as drop in feet divided by length in feet over which the drop takes place. The use of S defined thus in formula (7) involves certain assumptions that are approximations and can lead to error in some cases. It is assumed that the total energy gradient, or loss of head due to friction per unit length of channel, is equal to the slope of the channel bed, and to the slope of the water surface. This assumption is rarely, if ever true in natural rivers and streams, and if it is suspected that it is not approximately so, the reader should be prepared to accept the fact that a solution is not possible without more detailed study.

It should be made clear that the C in Chezy's formula is not a constant. Later experimenters found that C was a function of R and a coefficient of roughness n, and various formulas have been devised to include these variables. The Manning formula is considered to be the simplest and therefore the most useful. It is expressed as follows:

$$V = \frac{1.486}{n} R^{2/3} S^{1/2} \qquad (8)$$

Values of n, the roughness coefficient, for various types of channels are listed below:

Type of conduit	n
Galvanized wrought-iron pipe	0.014
Vitrified sewer pipe	0.013
Concrete pipe	0.013
Wood stave pipe	0.011
Concrete channels	0.014
Rough rock cuts	0.040
Shallow gravel channels (similar to salmon spawning grounds)	0.025
Natural river channels with clean stones	0.030
Natural river channels with weeds and stones	0.050

Various other formulas have been developed, and these are listed and explained in most hydraulics handbooks and texts. If greater accuracy or a check by another formula is required, the reader is referred to the list of references at the end of this chapter.

This formula can also be used for solving problems involving flow of water in pipes. For a pipe flowing full, the hydraulic radius R is equal to the pipe diameter d divided by 4. The slope S becomes the loss of head h_f/l. Thus the formula becomes

$$V = \frac{1.486}{n} \left(\frac{d}{4}\right)^{2/3} \left(\frac{h_f}{l}\right)^{1/2} \qquad (9)$$

It will be noted that values of n for some common types of pipe have been included in the foregoing table for use in this formula. It is usual to solve the formula for head loss h_f by transposing the terms.

A.9 MEASUREMENT OF FLOWING WATER

As noted in the introduction to this section, one of the most useful phases of hydraulics to the fisheries worker is the measurement of water flow. In order to ensure that a hatchery is running at maximum efficiency, it is necessary to know the flow through the various pipes and valves, and through the rearing ponds. Control of the flow within known limits is important in operating a fishway, and a method of measuring the flow is an essential prerequisite. If the flow through a screen is known, the velocity of approach can be calculated, and its efficiency estimated. Similarly, the flow through the screen bypass must be known and controlled, in order to make best use of the water for returning fish to their natural migration route. As pointed out in the introduction, it is extremely important to know the variation in flow in any natural stream in which the fish populations are under study. Variation in flow in a natural stream is nearly always the most important factor in determining the population the stream can support, since it controls the stream volume, which in turn controls the volume of food and living space, and often the temperatures. Measurement of flows is extremely important in hydraulic model work, where it is important to know as accurately as possible the velocities to be expected in the prototype under the same flow conditions. Hydraulic models are useful in solving a number of different kinds of fisheries problems, as was noted in Chapter 3 of this text. It is essential, therefore, that the reader realize the importance of measurements of water flow, even though in many cases the methods of making these measurements are beyond the scope of this text.

In some of the foregoing sections it has been noted that various devices such as orifices have been used for measuring flow. In this section it is proposed to group these methods according to their application in fisheries work and explain them in more detail so that they will be of use to the reader.

A.10 MEASUREMENT OF FLOW IN PIPES

Flow in pipes is encountered wherever water is used for fish cultural purposes. Often the quantities used are small, and can be measured in gallons per minute rather than cubic feet per second. The simplest method for measuring small flows such as those discharging into hatchery troughs or ponds, is to measure the length of time taken to fill a container of known dimensions. A 5-gal pail held under a pipe discharging into a hatchery pond, besides being the simplest method of measuring, is probably more accurate than many other of the methods that have been devised.

However, with larger flows it is not convenient to measure the flow in this way, and other methods are commonly used. One of the best-known methods is use of the Venturi meter, named after an Italian who devised it in the 18th century. It consists of a narrowed section (throat) of pipe connected to the main pipe by two cones of different lengths. The difference in pressure is measured between points at the entrance and at the throat, and the flow can be calculated from this difference. The whole meter is specially constructed so that inside diameters are extremely accurate. Certain precautions are taken in maintaining the correct proportions between the various dimensions of the meter, and factors such as temperature of the water may have to be taken into account. The reader who is interested in obtaining further details is referred to the texts listed in the references.

Other types of meters are used such as the orifice meter, which is similar to the Venturi meter, and various mechanical devices are commonly used in waterworks systems, data on which can be obtained from commercial suppliers. All these involve insertion of a special section containing the meter in the pipeline.

A.11 MEASUREMENT OF FLOWS BY WEIRS AND ORIFICES

Flow in hatchery ponds, fishways, and bypasses from screens can often be measured by a calibrated weir. If the weir used does not conform exactly to the shape and conditions used in weir tests for which data are available, the calibration must be done for the particular conditions encountered.

In a hatchery pond, with a weir at the inlet or outlet, an approximation of the flow can be made by use of the formula given in a previous section. If more accuracy is desired, a good hydraulics handbook would indicate whether the particular conditions encountered are reasonably well covered by an existing formula, with corrections for known conditions. As an alternative, however, the weir can be calibrated by measuring the flow using some other means such as filling a container of known volume and plotting a curve of discharge against head on the weir. From this curve it will be possible to obtain discharges by measuring the head on a staff gauge placed near the weir.

It is particularly important to know the discharge to be expected through fishways at dams, where the water is often valuable for hydroelectric power or other purposes. Assuming that the fishway is a pool and weir type, it probably will be possible to apply a weir formula with necessary corrections and determine the flow accurately. In many cases the formula given in this text will be sufficiently accurate. Care should be taken to make allowance for flow through any submerged ports in the baffles, since these are orifices that will carry flow in addition to the weirs in accordance with the principles outlined previously in this text. In addition, notched baffles will complicate the calculation of flows, but if due allowance is made, such as considering each section of weir crest at the same elevation separately, a sufficiently accurate result should be obtained. If a fishway is of any kind other than the pool and weir, it is unlikely that measurement of flow by weir would be practicable. However, some types of fish locks, such as the Borland, incorporate entrance or exit pools that are either filled or emptied by flow over weirs, which offer a ready means of calculating the total discharge.

Fishways with submerged orifices only and with no flow over the baffles, can be dealt with on the basis of the orifice principles given. However, in this case the coefficient of discharge becomes a very important factor. This coefficient can vary widely, depending on the form of the orifice. An orifice that is a rectangular hole in a thin plate will have a coefficient considerably different from one that incorporates a section of pipe that is, say, six times the baffle thickness in length. Once again, if the existing conditions simulate those for which data exist, the flow can be easily calculated; but if they do not, the particular orifice may have to be calibrated in a laboratory or in the field to determine the coefficient to be used.

Flow in fishways, such as the Denil type and the vertical slot type (or Hell's Gate), presents a special problem. Both types may have to be calibrated by means of a model in a laboratory to determine flows for the given conditions. Some work has been done on models of Denil fishways, but in practice the design is usually varied so much, in an effort to economize and simplify construction, that any work done previously usually cannot be depended on. Unless a fishway of this type duplicated a design that had already been tested, it should be calibrated in a laboratory for best results. Considerable work has been done on the vertical slot fishway in British Columbia and Washington State, and the coefficient of discharge has been fairly well defined for a number of different baffle sizes and designs. It is of interest to note that the vertical slot of this fishway does not conform to the accepted definition of an orifice, because the top of the slot is open. It could be regarded as a very special form of notched weir, particularly if a sill is used at the base, but because of its complete submergence and particular shape, it is much simpler in practice to regard it as an orifice.

One other location where the fisheries workers might encounter an opportunity for measuring flow by means of a weir is in the bypass from screens, particularly in installations in irrigation ditches. It is common practice at irrigation screens to lead the bypass water away from the ends of a screen by passing it over a weir. The water then discharges into a well and back to the river by means of a pipe or flume. The purpose of the weir (which is usually an adjustable set of stoplogs) is to control the outflow and at the same time to provide a head drop or velocity section over which the small fish cannot return once they have been diverted from the screen. Because water for irrigation is extremely valuable in arid areas, it is essential that the bypass flow be carefully controlled to prevent waste. Knowledge of the flow at all times is a prerequisite for good control, and this knowledge is easily obtained if the outflow is over a simple weir. The flow may be small enough to permit calibration of the weir by the method suggested previously for hatchery ponds, but in most cases the weir formula given will yield sufficiently accurate results.

A.12 MEASUREMENT OF FLOW IN NATURAL STREAMS

The flow in natural streams is usually determined by measuring A and V for use in the formula $Q = AV$. The value of A can be determined by taking depth measurements or soundings at convenient intervals across the stream and totaling the areas of the individual sections across the river.

The determination of V is more difficult. If only a rough measurement is desired, V might be determined by timing a surface float, such as a stick or chip of wood, over a known distance. It must be kept in mind, however, that these will be subject to extraneous influences such as wind, surface eddies, etc., which will reduce the accuracy. Specially made floats, such as a section of lightweight pipe weighted at one end so that it floats in a vertical position with only a small portion visible, will reduce these influences. If the velocity is obtained at or near the surface, multiplying by a coefficient of 0.85 at ordinary stages, and 0.9 to 0.95 at flood stages, will determine the mean velocity in the vertical section measured. If a long tube float is used that extends to within 0.9 of the depth of the stream, the velocity obtained will be close to the mean velocity for the vertical section.

In measuring stream discharges it is most convenient to divide the width of the stream into 5- or 10-ft strips or subareas, measuring the area of each strip and determining the mean velocity for each. The value of Q is then calculated for each

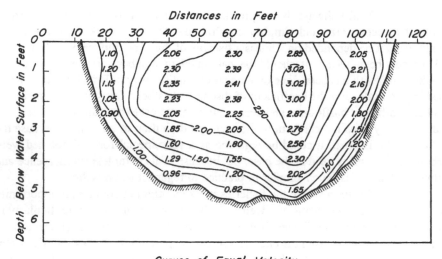

Figure A.3 Cross section of a stream showing velocity distribution. (From
Breed, C.B. and G.L. Howmer, *Principles and Practice of Surveying*,
Vol. 2, John Wiley & Sons. With permission.)

strip, and the sum of the discharges for all the strips represents the discharge of the
stream.

In determining the mean velocity more accurately in any vertical plane, experience has indicated that several methods are possible which entail a minimum of effort
but yield satisfactory results for practical purposes. The velocity distribution in a
typical stream is shown in Figure A.3. The curve of velocities on a vertical in a cross
section approximates a parabola. It has been found that an assumption can be made
that the curve is a parabola with sufficient accuracy for practical purposes. Therefore
by measuring the velocity at 0.2 and 0.8 of the depth, the mean of these two velocities
is equal to the mean velocity (from the parabolic curve, which gives a mean of all
ordinates by averaging the ordinates at 0.2114 and 0.7886). This method is widely
used in stream gauging and has been found from experiment to give virtually no error
directly attributable to the assumption. A second method sometimes used is to
measure the velocity only at 0.6 of the depth, which is close to the mean on the
parabolic curve. However, while this reduces the number of measurements necessary, it has been found to produce a small error in most results.

In the more accurate measurements utilizing the foregoing theory, a current
meter is used to determine the velocities at the various depths. Many of these are
available commercially, the common ones working on the principle of a small wheel
or propeller turned by the current. The turning of the wheel activates an electrical
contact, and the number of contacts is counted by an observer in a measured length
of time. This number varies with the velocity, so that a rating curve can be found for
each meter relating the number of revolutions to the velocity. Current meters are
normally designed to use by wading in shallow streams or by lowering from a bridge
or cable car.

DEPARTMENT OF FISHERIES — CANADA
CURRENT METER NOTES

Date _May 27,_ 1960 A.M.   Stream _Big Qualicum River_
Party _L.O. Scallon_   Locality _Near Horne Lake_
Meter № _56/794_. Gauge Height, Begins _2.65_ Ends _2.65_ Mean _2.65_
Total Area _83.00_ Mean Velocity _2.20_   Discharge _182.9_

Dist. from Point	Depth	Depth of Obsv'n	Revs	Time in Seconds	At Point	Mean Vertical	Mean Section	Area	Mean Depth Section	Width of Section	Dischge
0	0.2	0	0	0	0	0					
5	1.2	.7	20	57	0.81	0.81	0.41	3.50	0.70	5	1.44
10	1.1	.7	40	53	1.71	1.71	1.26	5.75	1.15	5	7.24
15	1.0	.6	70	55	2.86	2.86	2.28	5.25	1.05	5	11.97
20	1.7	1.0	60	54	2.51	2.51	2.68	6.75	1.35	5	18.09
25	1.6	1.0	60	51	2.66	2.66	2.58	8.25	1.65	5	21.28
30	2.0	0.4	60	46	2.94						
		1.6	80	70	2.57	2.76	2.71	9.00	1.80	5	24.39
35	2.2	0.4	60	54	2.51						
		1.8	60	60	2.26	2.38	2.57	10.50	2.1	5	26.98
40	2.3	0.5	80	61	2.95						
		1.8	80	70	2.57	2.76	2.57	11.25	2.25	5	28.91
45	2.0	0.4	50	43	2.63						
		1.6	50	54	2.10	2.36	2.56	10.75	2.15	5	23.11
50	1.4	0.8	30	43	1.58	1.58	1.97	8.50	1.70	5	16.74
55	0	0	0	0	0	0	0.79	3.50	0.70	5	2.76
Total											182.91

Figure A.4 Form used by the Department of Fisheries of Canada for recording data used in determining stream flow.

Figure A.4 shows a form used for stream gauging by the Department of Fisheries of Canada. In order to obtain a continuous record of stream flow it is necessary to know the discharge at least daily. Rather than repeat the velocity and cross-sectional area measurements each day, which would entail an enormous amount of work, it is usual to set a gauge in the stream and read only the elevation of the water surface daily. This can be done by an untrained observer or, under certain conditions, by a recording instrument. The gauge readings are plotted on a graph opposite a number of complete stream flow measurements for corresponding levels, and a curve is plotted, which is known as the *rating curve* for this section of stream. The rating curve must include stream flow measurements over the complete range of flows expected, as extrapolation of the curve is risky. It must also be checked at intervals if the reach of river in which it is established has an unstable bed, because this will cause the curve to change. From this curve it is possible to determine the flow corresponding to each gauge reading. It has been found convenient in fisheries conservation work to plot daily flows on a graph. Figure 2.7 shows a typical hydrograph of daily flows plotted on logarithmic paper. This shows the low flows with greater clarity, which is often helpful in fisheries work.

The setting up of a gauging station has been found from experience to involve a number of considerations that are beyond the scope of this text. These can be found in the references listed or in most good handbooks on hydraulics. It should be noted that Canada and the U.S., and probably in most other countries, the measurement of stream flow is done on a national basis by federal government agencies. These agencies are usually prepared to measure desired streams on request at cost, and since they are staffed with experienced personnel and are equipped, their services should be utilized whenever possible. However, there are always times when a measurement is needed in a hurry or is not available through the normal channels, and it is hoped the foregoing section will be of help in such cases.

Before concluding the section on stream flow measurement, it should be pointed out that the flow in very small streams can be measured accurately by a weir. They are comparatively easy to construct and will give a more accurate measurement of discharge than a point velocity method on small shallow streams. The use of the weir formula given in a previous section and with reasonable precautions to ensure accuracy, as already outlined, should give satisfactory results.

A.13 ADDITIONAL READING

Armco Drainage & Metal Products, 1946. *Handbook of Water Control,* Lederer, Street, Zeus Co., Berkeley, CA. 548 pp.

Davis, R.E. and F.S. Foote, 1940. *Surveying, Theory and Practice,* McGraw-Hill, New York. 1003 pp.

King, H.W., 1939. *Handbook of Hydraulics,* McGraw-Hill, New York. 605 pp.

Russell, G.E., 1940. *Hydraulics,* Holt, New York. 443 pp.

APPENDIX B
GLOSSARY OF COMMON
NAMES OF FISH USED

Common name	Scientific name
Acipenserid	Family Acipenseridae (sturgeons)
Alewife	*Alosa pseudoharengus*
Anchovy	Family Engraulidae (anchovies)
Northern anchovy	*Engraulis mordax*
Ayu	*Plecoglossus altivelis*
Atlantic tomcod	*Microgadus tomcod*
Bass	Family Centrarchidae (sunfishes); Family Percichthyidae (temperate basses)
Australian bass	*Macquaria novemaculeata*
Largemouth bass	*Micropterus salmoides*
Smallmouth bass	*Micropterus dolomieu*
Striped bass	*Morone saxatilis*
Bleak	*Alburnus alburnus*
Bluegill	*Lepomis macrochirus*
Bream	Family Cyprinidae (carps and minnows): *Abramis* spp.
Bullhead	Family Ictaluridae (bullhead catfishes)
Black bullhead	*Amieurus melas*
Bullrout	*Notestes robusta*
Carp	Family Cyprinidae (carps and minnows)
Grass carp	*Ctenopharyngoden idella*
Common carp	*Cyprinus carpio*
Crucian carp	*Carassius carassius*
Catostomid	Family Catostomidae (suckers)
Catfish	Family Ictaluridae (bullhead catfishes); Family Siluridae (sheatfishes)
Channel catfish	*Ictalurus punctatus*
Giant catfish	*Pangasianodon gigas*
White catfish	*I. catus*
Characin	Family Characidae (characins)
Chub	Family Cyprinidae (carps and minnows): *Couesius plumbeus* (lake chub), *Mylocheilus caurinus* (peamouth), other species

Common name	Scientific name
Cisco	Family Salmonidae (trouts)
Least cisco	*Coregonus sardinella*
Crappie	Family Centrarchidae (sunfishes)
Black crappie	*Pomoxis nigromaculatus*
White crappie	*Pomoxis annularis*
Croaker	Family Sciaenidae (drums)
White croaker	*Genyonemus lineatus*
Curimbata	*Prochilodus platensis*
Cyprinid	Family Cyprinidae (carps and minnows)
Dace	Family Cyprinidae (carps and minnows): *Margariscus margarita* (pearl dace), *Phoxinus eos* (northern redbelly dace), other species
Dorado	*Salminus maxillosus*
Eel	Family Anguillidae (freshwater eels)
American eel	*Anguilla rostrata*
European eel	*Anguilla anguilla*
Japanese eel	*Anguilla japonica*
New Zealand longfin eel	*Anguilla dieffenbachii*
Shortfin eel	*Anguilla australis*
Goldfish	*Carassius auratus*
Grayling	Family Salmonidae (trouts)
Arctic grayling	*Thymallus arcticus*
Gudgeon	*Gobio gobio*
Herring	Family Clupeidae (herrings)
Atlantic herring	*Clupea harengus*
Caspian herring	*Clupea harengus*
Hilsa	*Tenualosa ilisha*
Ictalurid	Family Ictaluridae (bullhead catfishes)
Inconnu	*Stenodus leucichthys*
Kokanee (landlocked sockeye salmon)	*Oncorhynchus nerka*
Lamprey	Family Petromyzontidae (lampreys): *Ichthyomyzon* spp., *Lampetra* spp., other genera
Sea lamprey	*Petromyzon marinus*
Leporinus	*Leporinus obtusidens*
Ling (burbot)	*Lota lota*
Loach	Family Cobitidae
Spine loach	*Cobitis taenia*
Mullet	Family Mugilidae (mullets)
Freshwater mullet	*Myxus capensis*
Sand mullet	*Myxus elongatus*
Striped mullet	*Mugil cephalus*
Paddlefish	*Polyodon spathula*
Peamouth	*Mylocheilus caurinus*
Perch	Family Centropomidae (snooks); Family Embiotocidae (surfperches); Family Percichthyidae (temperate basses); Family Percidae (perches)
Barramundi perch	*Lates calcarifer*
Eurasian perch	*Perca fluviatilis*

Common name	Scientific name
Golden perch	*Macquaria ambigua*
Nile perch	*Lates niloticus*
Shiner perch	*Cymatogaster aggregata*
Yellow perch	*Perca flavescens*
White perch	*Morone americana*
Perchlet	Family Ambassidae
Yellow perchlet	*Priopidichthys marianus*
Pike	Family Esocidae (pikes)
Northern pike	*Esox lucius*
Pumpkinseed	*Lepomis gibbosus*
Queenfish	*Seriphus politus*
Rainbow fish	*Melanotaenia* sp.
Salmon	Family Salmonidae (trouts)
Atlantic salmon	*Salmo salar*
Chinook (king) salmon	*Oncorhynchus tshawytscha*
Chum salmon	*Oncorhynchus keta*
Coho (silver) salmon	*Oncorhynchus kisutch*
Pink salmon	*Oncorhynchus gorbuscha*
Pacific salmon	*Oncorhynchus* spp.
Sockeye salmon	*Oncorhynchus nerka*
Sauger	*Stizostedion canadense*
Sculpin	Family Cottidae (sculpins)
Mottled sculpin	*Cottus bairdi*
Shad	Family Clupeidae (herrings)
Allice shad	*Alosa alosa*
American shad	*Alsoa sapidissima*
Threadfin shad	*Dorosoma petenense*
Shiner	Family Cyprinidae (carps and minnows): *Notropis* spp.
Smelt	Family Osmeridae (smelts)
Rainbow smelt	*Osmerus mordax*
Squawfish	Family Cyprinidae (carps and minnows)
Northern squawfish	*Ptychocheilus oregonensis*
Stickleback	Family Gasterosteidae
Sturgeon	Family Acipenseridae (sturgeons): *Acipenser* spp., *Huso* spp., *Scaphirhynchus* spp.
Sucker	Family Catostomidae (suckers)
Longnose sucker	*Catostomus catostomus*
Mountain sucker	*Catostomus platyrhynchus*
White sucker	*Catostomus commersoni*
Sunfish	Family Centrarchidae (sunfishes)
Green sunfish	*Lepomis cyanellus*
Tench	*Tinca tinca*
Threadfin	Family Polynemidae: *Polynemus* spp.
Trout	Family Salmonidae (trouts)
Brook trout	*Salvelinus fontinalis*
Brown trout	*Salmo trutta*
Cutthroat trout	*Oncorhynchus clarki*
Dolly Varden	*Salvelinus malma*
Golden trout	*Oncorhynchus aguabonita*

Common name	Scientific name
Lake trout	*Salvelinus namaycush*
Rainbow trout	*Oncorhynchus mykiss*
Sea trout (anadromous brown trout)	*Salmo trutta*
Steelhead (anadromous rainbow trout)	*Oncorhynchus mykiss*
Walleye	*Stizostedion vitreum*
Wels (sheatfish)	*Silurus glanis*
Whitefish	Family Salmonidae (trouts): *Coregonus* spp, *Prosopium* spp.
Mountain whitefish	*Prosopium williamsoni*

INDEX

9 780367 449261